Lecture Notes in Mobility

Series Editor

Gereon Meyer, VDI/VDE Innovation und Technik GmbH, Berlin, Germany

Lecture Notes in Mobility (LNMOB) is a book series that reports on the latest advances in research, development and innovations for the intelligent, connected and sustainable transportation systems of the future, comprising e.g.:

- Electric and hybrid vehicles
- Energy-efficient vehicles
- Alternative and optimized powertrains
- Vehicle automation and driver assistance
- Clean and intelligent transportation systems
- Interfaces with transportation networks and power grids
- Mobility services
- Business models
- Public policy

LNMOB publishes edited conference proceedings, contributed volumes and monographs that present firsthand information on cutting-edge research results and pioneering innovations as well as new perspectives on classical fields, while maintaining Springer's high standards of excellence. Also considered for publication are dissertations and other related material of exceptionally high quality and interest. The subject matter should be original and timely, reporting on the latest developments in all areas covered by this series. Concerns about climate change, air quality and energy security have sparked a paradigm shift regarding the transportation of passengers and goods. Vehicle technologies are now changing radically towards higher energy efficiency and the use of alternative fuels. At the same time, transportation is increasingly managed in a smart manner, connecting a multitude of modes and making use of a variety of energy sources. Moreover, vehicles are becoming more and more assistive and autonomous, which will lead to even higher efficiency, fewer fatalities and more convenience. These developments are to be supported by novel services, processes and value creation, as well as by suitable public policies. LNMOB identifies the most promising conceptual work, groundbreaking findings and innovations from the industrial engineers and academic experts in this field. The overall aim of this series is the exchange of ideas to strengthen the scientific community, to create new value chains and to accelerate the move towards sustainable transportation systems (road, rail, water- and airborne) around the globe. Public research funding programs in Europe, America and Asia represent an important source of content. The selection of work for publication is done under the supervision of an international scientific advisory board that brings together some of the most influential technology leaders and scholars of our time. The target audience primarily consists of industry professionals and academic researchers working at the forefront of their fields, but the book series will also be of interest to advanced level and PhD students.

More information about this series at http://www.springer.com/series/11573

Gereon Meyer · Sven Beiker
Editors

Road Vehicle Automation 7

Springer

Editors
Gereon Meyer
VDI/VDE Innovation + Technik GmbH
Berlin, Germany

Sven Beiker
Stanford University
Palo Alto, CA, USA

ISSN 2196-5544 ISSN 2196-5552 (electronic)
Lecture Notes in Mobility
ISBN 978-3-030-52839-3 ISBN 978-3-030-52840-9 (eBook)
https://doi.org/10.1007/978-3-030-52840-9

This Springer imprint is published by the registered company Springer Nature Switzerland AG
The registered company address is: Gewerbestrasse 11, 6330 Cham, Switzerland

Preface

When we started editing the Road Vehicle Automation books seven years ago, we were confident that the trend towards automated and connected driving would fundamentally change the idea of the automobile in the years to come. Rather than a transport machine that required steady human steering and control in order to avoid any harm to its driver, the occupants, and other road users, it would become a super safe and convenient autonomously moving shared space that mindfully adapts to the wishes of its users and to the situation around it. With the COVID-19 pandemic, it seems to be all different and is leaving room for asking whether we are still on the right track: transportation demand has dropped almost completely for weeks and is recovering just slowly. Both public and individual transport means are affected, and the promises of automated road mobility appear to be small compared to the desire for a readily available, trustworthily hygienic, and health-protecting way of getting from A to B. All of a sudden, the individually owned and controlled car seems to revive.

Actually, we believe that there is still a lot of potential in connected and automated road mobility, probably even more than ever. Automated vehicles can be seen as integral to mitigate the impact of the pandemic and to be an important part of our lives going forward. A self-driving and, maybe, self-sanitizing vehicle, e.g. could provide a kind of protective mobility shell for the vulnerable society members that would be superior to any kind of public transport in terms of infection safety. It could even put services on wheels which so far would have required people to travel to places where they risk being exposed to the virus. Or, as an autonomous delivery robot, it could offer a contactless provision of goods. Therefore, rather than lamenting about the adverse effects of the crisis, we should try to see the opportunities in it. An important additional benefit in this regard is the potential of automated vehicles to support a more efficient road traffic that could imply less fuel consumption, and thus, cleaner air and a more stable world climate.

The path promoting Road Vehicle Automation thus is more promising than ever. At the same time, though, one has to admit that significant progress is still needed to exploit its potential to the fullest: Ubiquitous self-driving functionality will require a number of breakthroughs in technology, such as environment perception systems, merging advanced sensing with artificial intelligence, virtual test fields for accelerated safety validation, and last but not least the complete and efficient integration of cars into the infrastructures for data, energy, and communication. At the same time, many non-technical issues, new and old, remain open for research and debate. One definitely should discuss, e.g. how the promises of self-driving shuttles in view of the COVID-19 pandemic can be realized in a way that is complementing public transport rather than cannibalizing it.

In this context, the Road Vehicle Automation books are playing an important role, as they have been documenting the progress in technical, economical, legal, societal, and human factors related matters of the development and implementation of automated road mobility from all around the globe for seven years now. A compendium like this surely is of high relevance in view of the need to exchange knowledge, cooperate, and jointly find new solutions for accelerating innovation. This is also, and in particular, true in response to the pandemic crisis, which is changing the world, seemingly in a dramatic way and at a dramatic rate. The contents may even show directions for public and private investment in research funding and recovery measures.

Therefore, we are fortunate that a number of plenary speakers and breakout session owners of the Automated Vehicles Symposium (AVS) 2019 have agreed to present their recent achievements and insights into the innovation paths to the future of connected and automated road mobility for this book. We are also indebted to the organizers of AVS on behalf of the Transportation Research Board (TRB), Jane Lappin, Valerie Shuman, and Steve Shladover, for their kind support of this publication. Special thanks go to the Applied Sciences and Engineering editorial team at Springer, to the Association of Automated Vehicle Systems International (AUVSI), and to colleagues at VDI/VDE-IT for assisting us in turning this publication into reality despite the difficulties due to the pandemic. It shows that all the hard and diverse work on Road Vehicle Automation might have led to this point enabling and creating much of our societies, economies, and administrations for the future. And, we sincerely hope that it will be a good future, filled with health, fairness, and prosperity.

In that sense, even though the Automated Vehicles Symposium 2020 will be held as a virtual event, we hope to interact with many of our much-appreciated contributors as well as the entire community to discuss this very publication and to plan the next sequel, Road Vehicle Automation 8. For the time being, we hope this book will find its way to you as a reader being safe and healthy.

June 2020

Gereon Meyer
Sven Beiker

Contents

Introduction: The Automated Vehicles Symposium 2019

Steven E. Shladover[1(✉)], Jane Lappin[2], and Valerie Shuman[3]

[1] University of California PATH Program, 1357 South 46th Street, Building 452, Richmond, CA 94804, USA
steve@path.berkeley.edu
[2] TRB Vehicle-Highway Automation Committee, Washington, DC, USA
JaneLappin@gmail.com
[3] Shuman Consulting Group, LLC, Skokie, IL, USA
vs@shumangroupllc.com

Abstract. The 2019 Automated Vehicles Symposium followed a similar pattern to its predecessors, but with a change of location from California to Orlando, Florida, to offer a different local perspective. The plenary and poster presentations and breakout discussions continued to provide the meeting participants with the most up-to-date and authoritative information about the current international state of development and deployment of road vehicle automation systems, continuing to make this the essential meeting for industry, government and research practitioners in the field.

Keywords: Road vehicle automation · Road transport automation · Automated vehicles · Autonomous vehicles · Self-driving vehicles

1 Overview

The 2019 Automated Vehicles Symposium was organized and produced through a partnership between the National Academies of Science, Engineering and Medicine (NASEM) Transportation Research Board (TRB) and the Association for Unmanned Vehicle Systems International (AUVSI), continuing the pattern established in the five preceding years. This meeting was organized to serve their constituencies' interests in understanding the impacts, benefits, challenges and risks associated with increasingly automated road vehicles and the environments in which they operate. It brought together key government, industry and academic experts from around the world with the goal of identifying opportunities and challenges and advancing automated vehicle (AV) and highly automated driving (HAD) research across a range of disciplines.

The symposium took place over five days, 15–19 July, 2019 with four days of core activities and ancillary sessions on the first and fifth mornings. The three morning plenary sessions included presentations from the public sector, automakers and suppliers and research institutes and the afternoons were devoted to thirty-five breakout sessions for deeper investigation and discussion of selected topics. A reception with a poster session in the exhibit hall followed the close of the breakout sessions on both

G. Meyer and S. Beiker (Eds.): AVS 2019, LNMOB, pp. 1–9, 2020.
https://doi.org/10.1007/978-3-030-52840-9_1

Tuesday and Wednesday afternoons, with 33 posters presented in each of the poster sessions..

The breakout sessions were each organized by committees of volunteers to address a wide range of topics. These were clustered into four thematic tracks to make it easier for attendees to identify the sessions of strongest interest to them:

- Policy and Planning
- Users and Human Factors
- Operations and Applications
- Technology.

The symposium also involved several related meetings that occurred before and following the main meeting:

- U.S. DOT Listening Session
- National Cooperative Highway Research Program panel 20-102, sponsoring research on impacts of connected vehicles and automated vehicles on state and local transportation agencies
- SAE On-Road Automated Driving (ORAD) Standards Committee meeting
- Meeting of the TRB Committee on Automated Transit Systems (AP040)
- Meeting of the TRB Forum on Preparing for Automated Vehicles and Shared Mobility Systems
- U.S. – Japan – EU Trilateral Working Group on Automation in Road Transportation

In keeping with TRB practice, the plenary and breakout sessions were planned and produced by volunteers whose expertise and work informed the content of the sessions. In keeping with AUVSI practice, the production of the symposium was professionally managed by dedicated conference and logistics managers. The AVS19 Executive Committee reflected this mix of the two organizations:

David Agnew, Dataspeed, Inc. (and AUVSI board member), Richard Bishop, AUVSI subject matter expert on automation; Allison Cullin, Amazon, Richard Cunard, Senior Program Officer, Traffic and Operations Engineer, TRB; Steven Dellenback, Southwest Research Institute, Kevin Dopart, U.S. DOT Intelligent Transportation Systems Joint Program Office, Jane Lappin, Toyota Research Institute and TRB Vehicle-Highway Automation Committee Chair; William Malley, Perkins Coie, Steven Shladover, University of California PATH Program (and former chair of the TRB Vehicle-Highway Automation Committee); Valerie Shuman, Shuman Consulting Group, LLC and Chair, TRB CORVA Subcommittee, Edward Straub SAE, and Lindsay Voss, Senior Program Development Manager, AUVSI.

2 Symposium Attendees

Almost 1400 registrants participated in the symposium, somewhat below the attendance the preceding year based on the significantly longer distance from the hotbed of automation development activity in Silicon Valley. Attendees represented a wide range of organizations from government and industry to the academic-, public-, and private-sector research communities. One of the strengths of the meeting was the breadth of interests represented, including industry, public agencies and academic/research organizations. The road vehicle industry was well-represented with attendees from Original Equipment Manufacturers (OEMs) and their suppliers.

These participants represented disciplines ranging from engineering to psychology to law. Twenty-four countries (representing the 15% of the meeting participants who come from outside the U.S.) and twenty-four U.S. states were represented among the meeting participants. The largest delegation from outside the U.S. came from Japan, with 36 participants, while Canada, South Korea and Germany all had more than ten participants. Florida, as the host state, had the largest number of attendees from within the U.S., followed by the national capital region (DC, Maryland, and Virginia), California, and Michigan. The overall distribution was not quite as geographically diverse, both nationally and internationally, as it had been in previous years when the meeting was in San Francisco.

3 Keynote Talks

AVS 19 got off to a great start with a keynote talk by Chris Urmson, who formerly headed the Google Self-Driving Car project and is now the co-founder and CEO of Aurora Innovations. He began with a review of his history with automated driving, beginning with the DARPA Challenges, and then focused on the safety challenges that need to be mastered. He proposed that the acceptable risk baseline for comparison should be the existing NHTSA crash safety statistics, which can be expressed in terms of the rate of occurrence of crashes of varying levels of severity, plotted on a logarithmic scale. The risks associated with any specific ADS implementation can be allocated based on quantitative estimates of the perception and planning errors. Machine learning will be necessary for some functions because of the complexity of the driving environment and its hazards, but because its safety cannot be assured directly, virtual "guardrails" need to be implemented to limit the actions machine learning systems can command.

Chris noted that ensuring safety during the development and testing processes will be essential for building public trust, and he explained some of the strategies that Aurora has implemented to ensure safety during their testing. An organization-wide safety culture is important to establish. They also use virtual testing in simulations to reduce the amount of on-road testing needed to refine their system, limiting the exposure of the public to testing risks.

AVS19 also had the benefit of two keynote addresses from the U.S. Department of Transportation, from Federal Highway Administrator Nicole Nason and Federal Motor Carrier Safety Administrator Raymond P. Martinez.

FHWA Administrator Nicole Nason opened her remarks by conveying USDOT Secretary Elaine L. Chao's greetings to the symposium. She cited the USDOT's "Preparing for the Future of Transportation: Automated Vehicles 3.0," which provides guidance for managing safety risks and clarifying government roles. Ms. Nason expressed the Department's commitment to the 5.9 GHz "safety band" for reducing crashes, injuries, fatalities, and overall traffic congestion.

As FHWA Administrator, she highlighted the agency's recent activities leading the National Dialogue on Highway Automation, reviewing the Manual on Uniform Traffic Control Devices, launching the "Data For Automated Vehicle Initiative," field-testing truck platoons to assess their safety, efficiency and mobility impacts, and leading the "Cooperative Automation Research Mobility Application" (CARMA) to accelerate understanding of the benefits of cooperative automation by testing shared maneuvers such as vehicle platooning, speed harmonization, cooperative lane change and merge functions, and coordination of signalized intersections. Finally, Ms. Nason described the Department's newly established Non-Traditional and Emerging Transportation Technology (NETT) Council, which provides coordination across the operating agencies for the deployment of cutting-edge technologies.

FMCSA Administrator Raymond Martinez opened his remarks with an invitation to attend the Multi-Modal Automated Vehicle Listening session to be held that afternoon. As the lead Federal agency responsible for regulating commercial motor vehicles (CMVs) and the nearly 4.7 million active holders of commercial drivers' licenses, he stressed the agency's role in bringing lifesaving, assistive technologies to the market quickly. In May 2018, the Agency released a Request for Comments on FMCSA safety regulations that may inadvertently pose barriers to the safe testing and deployment of ADS technologies on public roads. Under existing regulations regarding operating ADS-equipped CMVs, FMCSA will no longer assume that a CMV driver is always a human, or that a human is necessarily present onboard a commercial vehicle during operation.

Moving forward, the Agency is seeking input on testing vehicles with automated driving technologies. FMCSA issued an Advance Notice of Proposed Rulemaking seeking public comment to better understand how changes to its rules can account for significant safety differences between human operators and automated driving systems. The Agency is also working to identify key safety technologies to better understand the impact of automated vehicles on the nation's freight system. FMCSA is working with FHWA on truck platooning exercises at the U.S. Army testing facility at Aberdeen, with the Maritime Administration to explore truck automation improvements at intermodal port facilities, and with NTSB to "Increase Implementation of Collision Avoidance Systems in All New Highway Vehicles". Finally, FMCSA is planning a four-year project to work with motor carrier industry associations to promote the benefits of collision avoidance systems through education, outreach, and training.

4 Plenary Panel Sessions

AVS19 devoted more of the plenary program time than previous AVS meetings to panel discussion sessions featuring groups of speakers responding to questions from the moderator and interacting with each other, to break up the sequence of formal presentations.

Steven Shladover chaired a plenary panel session on stakeholder perspectives on safety assurance for automated driving systems, with panelists Steve Gehring from Global Automakers, David Kidd from the Insurance Institute for Highway Safety, David Yang from the AAA Foundation for Traffic Safety and Peter Norton from the

University of Virginia. This session was focused on the institutional side of safety assurance, and in particular how to provide convincing evidence to the public regarding the safety of automated driving systems in the absence of benchmark government regulations or standards. The speakers discussed the value of sharing data and experiences among the developers of the automated systems to accelerate safety improvements, so that safety is not treated as a competitive issue. Information provided to consumers by independent third-party evaluators and through rating systems such as NCAP can help to educate them about the safety of different systems. Developers of automation systems were also encouraged to interact with regulators during the development process to get feedback and advice, rather than waiting until a product release.

Kelley Coyner chaired a plenary panel session on advancing urban mobility using vehicle automation systems, with panelists Lauren Isaac from EasyMile, Henry Greenidge from Cruise Automation and Ryan Jacobs from Aptiv Autonomous Mobility, who discussed the practical issues involved in testing their automated driving systems in specific urban environments.

Richard Bishop chaired a plenary panel session about early deployments of connected and automated heavy duty trucking, with panelists Christian Haas from Fresenius University of Applied Sciences, Joshua Switkes from Peloton Technology, Chuck Price from TuSimple, Keith Brandis from Volvo Group, and Kelly Regal from the Federal Motor Carrier Safety Administration. The panelists explained the experiences they have had testing platooning systems on public roads at low levels of automation (Haas and Switkes) and developing systems with higher levels of automation (Price and Brandis). The Peloton operational testing of Level 1 platooning produced field test data about fuel savings and showing how infrequently cut-ins and hard braking occurred in real-world operations, while the Fresenius testing contributed to understanding of driver human factors in Level 2 platooning. TUSimple and Volvo emphasized their work on Level 4 automation of individual trucks, which was followed by spirited discussion about whether Level 4 automation would first be implemented on following trucks in platoons or individually driven trucks.

NHTSA organized a plenary panel session about the interactions between public safety personnel and highly automated vehicles, chaired by Jon Krohmer and Dee Williams. This panel featured Chief Derek Barrs from the Florida Highway Patrol, Dr. Hezedean A. Smith from the Orlando Fire Department, Al Prescott from Tesla, Andre Welch from Ford and Matt Schwall from Waymo. This provided an opportunity for the public safety officials to explain their needs and concerns to the automation system developers, and for the developers to explain what they are doing to facilitate the interactions of automated driving systems with public safety officers. There is a clear need for the system developers to provide more information to the public safety community about how their systems operate and to engage directly with public safety officials, particularly in training them how to interact with vehicles that operate without drivers for scenarios involving vehicle failures, criminal activities, passenger medical crises and other emergencies.

Bob Denaro chaired a panel session on investor perspectives on automated driving with panelists Jim Adler, TRI and Olaf Sakkers, Maniv Mobility. They noted that the competitive dynamics in this field are different from the traditional automotive model, driven less by competition among automotive OEMs than by competition between them and the new arrivals from the information technology industry. They also noted that the initial hype period has passed and we are already entering a period of consolidation, with investors becoming more cautious about how they choose to invest. Their interests are now more focused on developers of essential enabling technologies for automation than on "full-stack" automation system integrators (with the exception of those that can identify a narrowly defined application niche within a simplified operational design domain).

5 Plenary Presentations

The plenary presentations were primarily clustered in multi-speaker sessions, with opportunities for some interactive discussion with a moderator following the presentations, although a few presentations were done on a stand-alone basis. These presentations were generally targeted at addressing cross-cutting issues associated with road vehicle automation rather than focusing on specific technologies or development activities. The presentations are grouped here in broad thematic categories:

Technical and Policy Challenges in Safety Assurance:

- PEGASUS Results and Future Prospects - Dr. Lutz Eckstein, RWTH Aachen University
- Building Trust in AV Safety - Noah Zych, Uber Advanced Technologies Group
- Volvo Cars Automation Safety Assurance Framework - Trent Victor, Volvo Cars
- AV Safety Assurance Principles, Standards, and Best Practices - Ed Straub, SAE Office of Automation

User Attitudes and Interactions and Policy Considerations

- What the Public Really Thinks About Automated Vehicles: Evidence From Survey Research - Johanna Zmud, Texas A&M Transportation Institute
- Engaging the Public with Automated Vehicles - Jack Stilgoe, University College London
- Trust Built on Knowledge: Partners for Automated Vehicle Education - Kelly Nantel, National Safety Council
- What We've Learned After Providing 50,000 Self-Driving Rides to the Public - Jody Kelman, Lyft
- External Interfaces for Automated Vehicles from European Commission's interACT Project - Natasha Merat, University of Leeds
- Congressional Staff Perspectives on National Policies for Road Vehicle Automation - Cheri Pascoe, U.S. Senate Committee on Commerce, Science, and Transportation

Public Sector Activities on Road Vehicle Automation:

- European Policy and Initiatives Regarding Connected and Automated Driving - Tom Alkim, European Commission
- Putting the UK at the Forefront of the Self-Driving Future - Jess de Looy-Hyde, UK Centre for Connected and Autonomous Vehicles

- Japan's SIP-adus Program on Road Vehicle Automation - Ryota Shirato, Nissan Motor Co., Ltd.
- Developing End-to-End Regulation for Automated Vehicles in Australia - Marcus Burke, National Transport Commission
- Gearing up for Autonomous Mobility: Singapore's Approach - Chris Leck, Land Transport Authority
- U.S. Department of Transportation Automated Vehicle Research Activities – Finch Fulton, U.S. DOT

6 Breakout Sessions

The breakout sessions provided opportunities for more in-depth consideration of specific topic areas among groups of people with focused interests in those areas. With smaller groups, they could be more interactive than the large plenary sessions, providing ample opportunities for questions and answers and debates. The primary findings from each afternoon's breakout discussions were reported back to the plenary the following morning.

The breakout sessions covered a wide range of specialized topics relevant to automated driving to match the interests of different groups of meeting participants. The report-outs to the plenary session revealed some of the important lessons learned from the breakout discussions that have broad significance across the field. The majority of the breakout sessions covered a single afternoon, but a few of them extended to two afternoons (designated by (2) in the listing below).

Policy and Planning Sessions:

- Preparing for Automated Vehicles and Shared Mobility: The Existential Questions
- Working with Infrastructure Owner-Operators to Overcome Public Sector Institutional Barriers and Safely Implement Roadway Automation
- Ethical Algorithms in Autonomous Vehicles (2)
- Shark Tank III: Active Debate Regarding AVs Impact on Land Use; Resiliency; Congestion Pricing; High-Speed Rail
- Energy and Environmental Implications of Connected and Automated Vehicles: Trends in Industry, Research, Regulations and Policy
- AV-Readiness Planning in MPO Long-Range Transportation Plans
- Regulatory Policy for Automated Vehicles
- An AV Crash Happens: The Trial

Users and Human Factors Sessions:

- Not So Autonomous Cars: A Path to Consumers' Changing World
- Automated Vehicles and Vulnerable Road Users: A Focus on Under-Represented User Groups
- How Can Automation Improve Rural Accessibility and Mobility?
- HMI Design Strategies for Assisted Driving Automation
- Understanding Travel Behaviors in an Automated World

- Understanding the Value of External HMI in Communication of Intent by Automated Vehicles

Operations and Applications Sessions:

- New Innovations in Intelligent Intersection Management with Cooperative Automation
- Planning-Level Capacity Adjustments for CAVs: The Future is Now
- CAV Activities in Florida
- Trucking Automation: Deployment Challenges and Opportunities (2)
- Automated Vehicle (AV) Data – Who Has It? Who Wants It? What Format?
- Data for AV Integration
- Connected Infrastructure Systems Enabling Automated Vehicles in Smart Communities
- Gamechanger! Using Dedicated Lanes for Early AV Deployment
- Automation in Mobility: Where Are We and Where Do We Need To Go?
- Enabling Transportation Network: From Individual Vehicle Motion Control to Network Fleet Management
- Catching Up with Low-Speed Autonomous Shuttles

Technology Sessions:

- CARMA – Automated Vehicles Working Together
- Artificial Intelligence and Machine Learning in Infrastructure Readiness for Vehicle Automation
- New Simulation Tools for Training and Testing Automated Vehicles
- Reading the Road Ahead: Preparing Highway Infrastructure for ADAS and AVs
- Safety Assurance of Automated Driving (2)
- Spectrum Needs for Cooperative Automation
- At the End of the Road: Off-Road Automation
- Automated Vehicles Ecosystem End-to-End Cybersecurity
- Blockchain: Enabling Coordinated Autonomy
- Enabling Technologies – A Peek Under the Hood

7 General Cross-Cutting Observations

As the field of road vehicle automation has advanced and the level of knowledge of the issues has grown over the past several years, the areas of emphasis within the Automated Vehicles Symposium have shifted. In this most recent meeting, several general observations are worth noting:

- The trend from the last couple of years toward more realistic discussion of the scope and timing of Automated Driving System (ADS) deployments continued and broadened, with a wider range of participants shifting away from hype and toward more cautious predictions. This was most notable in terms of emphasizing that the initial deployments and those for the foreseeable future will be for narrowly defined operational design domains (ODDs) rather than being nationwide in scope.

- This caution extended into the discussions of public outreach and engagement with the media, elected officials and general public – tempering the outreach messages with more conservative predictions of capabilities and impacts, and also listening more carefully to public concerns as part of a dialogue rather than just trying to "sell" ADS to them.
- The early experiences of the public with advanced driver assistance systems (ADAS) and the initial prototype low-speed shuttle systems are conditioning public opinions about the prospects for longer-term ADS, generally in a positive direction, but sometimes negatively (when systems do not perform well). This emphasizes the importance of making sure that the performance of these early systems is improved before pushing toward higher levels of automation.
- There was broad recognition of the need for collaboration across countries and across companies because of the scope of the technical challenges involved and the resources needed to meet those challenges. This could be seen as a reflection of the recent widespread consolidations of efforts in partnerships occurring within the industry.
- The industry participants expressed more willingness than in the past to share experiences and some kinds of information, such as hazard scenarios, testing procedures and general lessons learned, and also to work together on development of relevant standards.
- There was widespread interest in finding common approaches to safety assurance for ADS development and certification for public use. This included common methods of assessment, criteria for acceptance and development approaches such as Safety of the Intended Functionality (SOTIF), which had barely been mentioned in prior years but was cited by multiple speakers this year.
- There was also considerable discussion about many data-related topics, including the data needed to support ADS development, the ownership of data, and data privacy and intellectual property issues.

Part I Public Sector Activities

Japan's SIP-Adus Program on Road Vehicle Automation

Ryota Shirato[1(✉)] and Seigo Kuzumaki[2]

[1] Nissan Motor Company, 560-2, Okatsukoku, Atsugi 243-0192, Japan
r-shirato@mail.nissan.co.jp
[2] Toyota Motor Corporation, 1, Toyota-Cho, Toyota 471-8572, Japan
seigo_kuzumaki@mail.toyota.co.jp

Abstract. SIP, the Cross-ministerial **S**trategic **I**nnovation **P**romotion program, was started in 2014 with 11 research projects, and those projects are proceeded by initiative of Council for Science, Technology and Innovation of Japanese government. SIP-adus, **a**utomated **d**riving system for **u**niversal **s**ervice, is one of the SIP projects and aims to realize innovation of the Automated Driving System through fundamental research to practical application and commercialization. The second phase started from 2018 and is planning Field Operational Test at Tokyo waterfront area in 2020. This paper introduces Society 5.0, Japan's proposal in the 5th Science and Technology Basic Plan, and an overview of the 1st and 2nd phase of SIP-adus and its challenges and contributions.

Keywords: Automated driving · Automated vehicles · Connected vehicles · Dynamic map · Field operational test · Traffic environmental information framework · SIP-adus · Japan

1 Introduction: Society 5.0

Japan has its particular challenges for digital transportation of manufacturing. Society 5.0 was proposed in the 5th Science and Technology Basic Plan as a future society that Japan should aspire to. It follows the hunting society, agricultural society, industrial society, and information society. Society 5.0 achieves a high degree of data convergence between cyberspace (virtual space) and physical space (real space). It leads economic advancement and solutions of social problems, which provides products and services to the people that need them at the time they are needed. Finally, a human-centered society in which anyone can enjoy a high quality of life with full of vigor will be realized (Fig. 1) [1]. In Society 5.0, new value can be generated through AI analysis of big data including sensor data from automobiles, real-time information on the weather, traffic, accommodations, food and drink, and personal history.

It is expected to optimally plan for travel, reduction of congestion and traffic accident, smooth transfer, and movement support for the elderly and physically challenged through the use of automated driving technology. Furthermore, these solutions will help reduction of CO2 emissions by public transportation (Fig. 2). SIP-adus aims to realize that innovative future society.

G. Meyer and S. Beiker (Eds.): AVS 2019, LNMOB, pp. 13–22, 2020.
https://doi.org/10.1007/978-3-030-52840-9_2

Fig. 1. Society 5.0

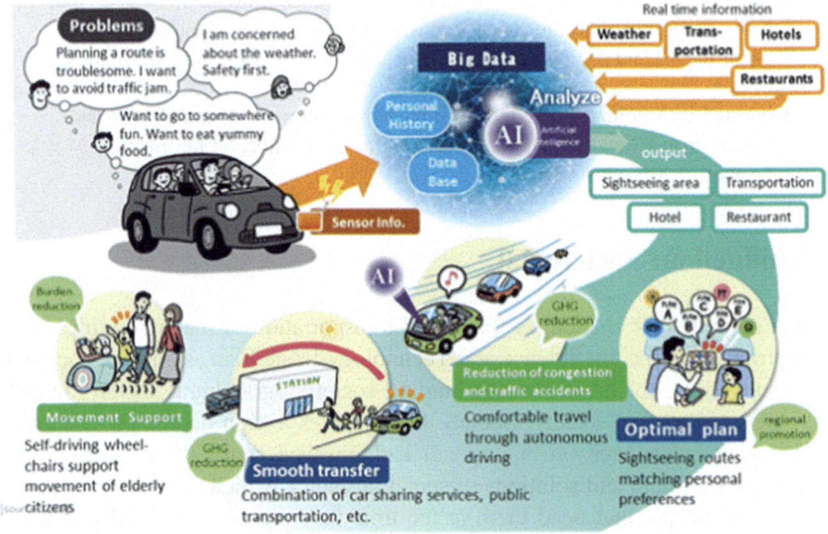

Fig. 2. Expected solutions of Society 5.0

2 Outline of SIP

SIP was started in 2014, and planned as 5 years project. It aims to realize Science, Technology and Innovation from fundamental research to practical and commercialization by cross-ministerial cooperation. By initiative of CSTI, Council for Science, Technology and Innovation, SIP program are proceeded. CSTI appointed the Program

Director and allocated the budget for each research theme. And it promotes the enhanced cross-ministerial cooperation and industry-academia-government collaboration (Fig. 3).

Fig. 3. Outline of SIP

SIP has the following features.

- Intensive R&D program
  Promote 5-years R&D
  1st phase: FY2014–FY2018
  2nd phase: FY2018–FY2022
  From fundamental research to practical and commercialization
- Promote cross-sector collaboration
  Enhancing cross-ministerial cooperation
  Promoting industry-academia-government collaboration
- Leadership and total Budget
  CSTI appointed Program Directors and allocates the budget for each research theme.

SIP started with 11 research projects in the area of Energy, Next-generation Infrastructures and Local Resources. The theme of Automated Driving System was selected from the beginning, and its objective is developing new system for more safe and convenient transportation. [2]

3 1st Phase of SIP-Adus (FY2014–FY2018)

3.1 Goal and Deployment Milestone

When starting up SIP-adus, it had a goal of reducing the number of traffic accident fatalities as the highest-priority. It also had the secondary goal of hastening the implementation of an automated driving system. Third goal was to realize a next-generation urban traffic system for 2020.

Figure 4 shows the JAMA, Japan Automobile Manufacture Association's, roadmap for the automated driving system. The longitudinal axis shows an application of automated driving systems, from simple areas such as highways to complex areas like general roads. The horizontal axis means a progress of systems from autonomous vehicle to connected automated vehicle with ITS.

SIP-adus supported to realize this scenario from the government point of view.

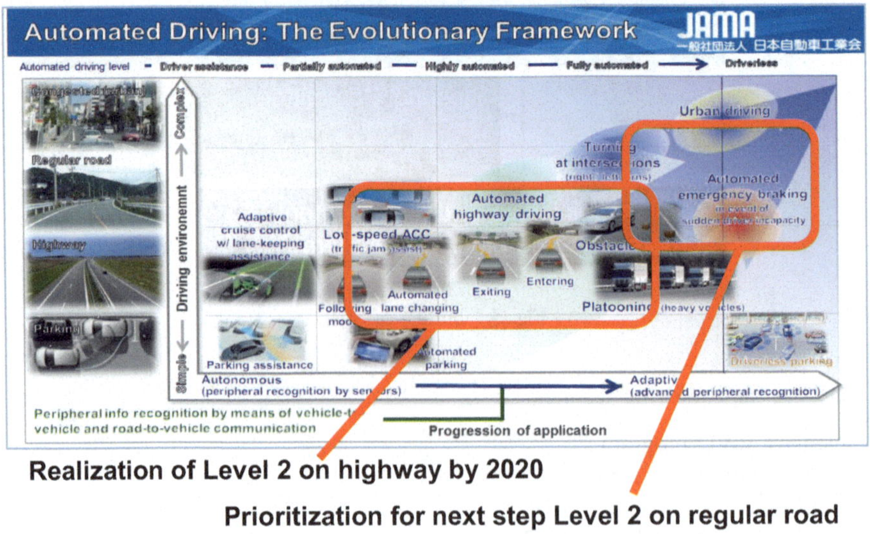

Fig. 4. JAMA's automated driving system roadmap.

3.2 Main Technology Domain

For automated driving systems, it is necessary to develop various technologies, such as high performance on-board sensors like cameras or radars for recognition and artificial intelligence for judgment. In SIP-adus, it was difficult to cover all relevant themes with the limited resources available. Therefore, it was discussed to classify the technology in cooperative field from all themes.

Figure 5 shows the main technology domain of 1st phase of SIP-adus. The letters in red indicate the cooperative themes, e.g. dynamic map, HMI, cyber security, and simulation.

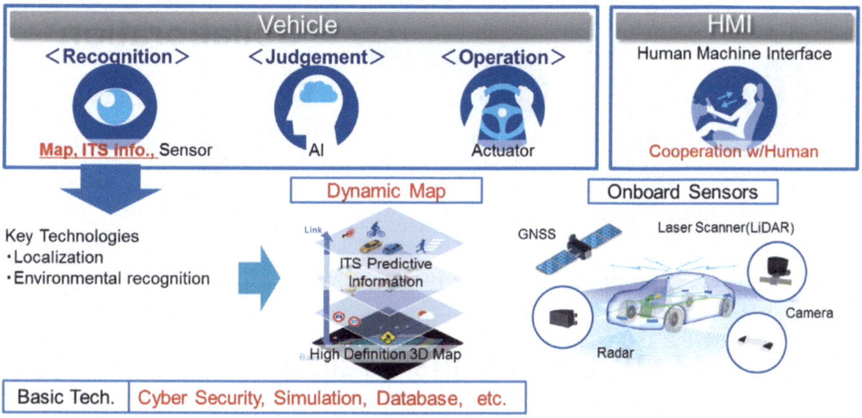

Fig. 5. Main technology domain of 1st phase of SIP-adus.

The automobile is an international product, so international harmonization and standardization is also very important. [3]

3.3 Output of 1st Phase of SIP-Adus

The aim of SIP is to realize innovation through fundamental research to practical application and commercialization by cross-ministerial cooperation and industry-academia-government collaboration. Healthy competition and cross-sectional cooperation are necessary for automated driving realization. So SIP-adus was tackling the issues with the R&D in cooperative field in SIP-adus.

One of the major output from first phase was digital infrastructure. What are necessary features for Automated Driving was discussed and the sample of precise 3D map data was created. More features are better, but increase cost.

As the result, Dynamic Map Planning, DMP, was established in 2016. Six map companies and nine automakers invested in the company. DMP was developing the methodologies of creating and maintaining a high-precision 3D map data for automated driving systems. In June 2017, the planning phase was completed, and DMP became a business enterprise, and name of DMP was changed to Dynamic Map Platform (Fig. 6).

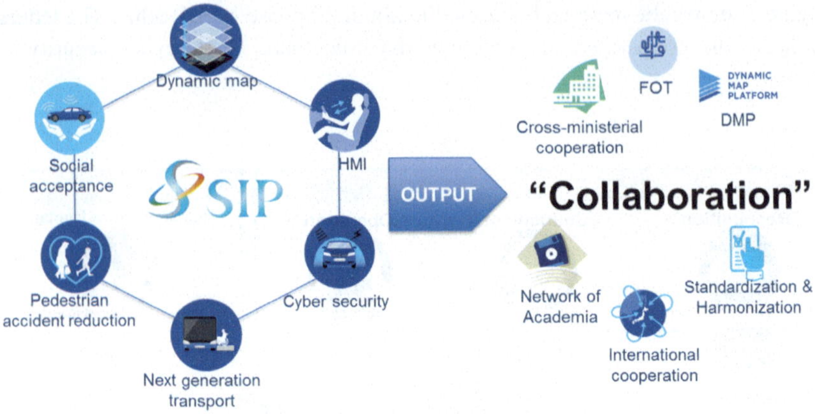

Fig. 6. Output of 1st phase of SIP-adus.

4 2nd Phase of SIP-Adus (FY2018–FY2022)

4.1 Overview

There are two possible approaches for automated driving development as shown in Fig. 7.

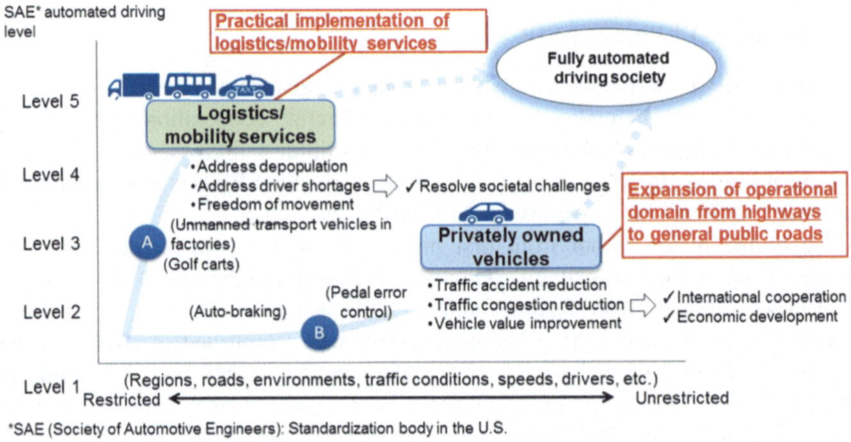

Fig. 7. Approaches for automated driving development.

The vertical axis is the level of automation according to the SAE definition, and the horizontal axis indicates the operational design domain (ODD). Left is a restricted condition, and the further right, the less restricted.

In its first phase, SIP-adus focused on the development of automated driving on highway for privately owned vehicles, so the second phase has to expand the operational domain from highway to public roads. Also, regarding logistic and mobility services, it is critical to implement automated driving service for solving social issues like mobility in rural area and shortage of truck and bus drivers.

SIP-adus aims to promote R&D overlooking from fundamental research to practical application and commercialization, so stakeholders of commercialization participate in the R&D phase. And it is also expected that mobility services will be commercialized smoothly at the end of the project.

The second phase has a strategy of developing milestones. Specifically, investment and business planning by private operators will be promoted by:

1) Taking full advantage of the Olympic and Paralympic Games Tokyo occasion
2) Conducting field operational tests (FOT) based on the plans of business operators and local government

4.2 Four Pillars of 2nd SIP-Adus

The second phase is composed of four pillars. First, FOT are created in order to supply opportunity for open discussion. For realization of automated driving systems, it is said that there are three barriers to overcome, which are technology, law and social acceptance. Regarding regulatory reform and rulemaking, all Japanese ministries are making efforts to solve those issues. So SIP-adus focuses on development of core technology and fostering of social acceptance as second and third pillars. The fourth pillar is international cooperation. SIP-adus aims to promote the joint research with overseas research institutions (Fig. 8).

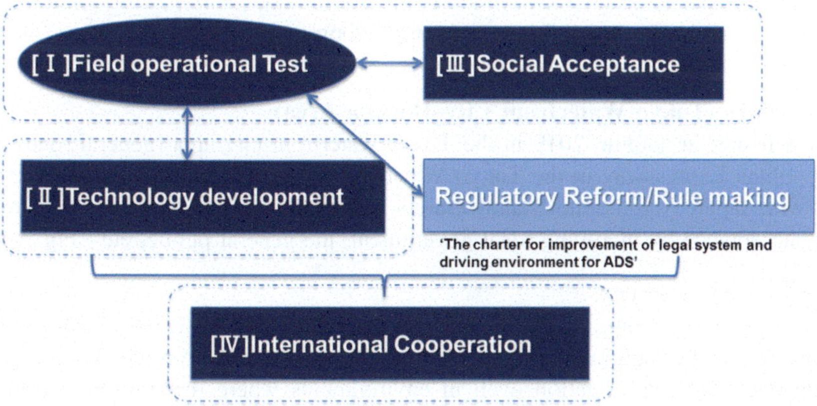

Fig. 8. Four pillars of 2nd SIP-adus

4.3 Major Activities

4.3.1 Traffic Environmental Information Framework

In the first phase of SIP, a basic initiative of dynamic map, and developing 3D precise static map data were established. In the second phase, it aims to develop and operate dynamic traffic data, such as traffic light information and merging area traffic information, and also planning to do Field Operational Testing, FOT, in the Tokyo water front area. Finally, it is expected to realize a cooperative automated driving society, and new data business creation (Fig. 9).

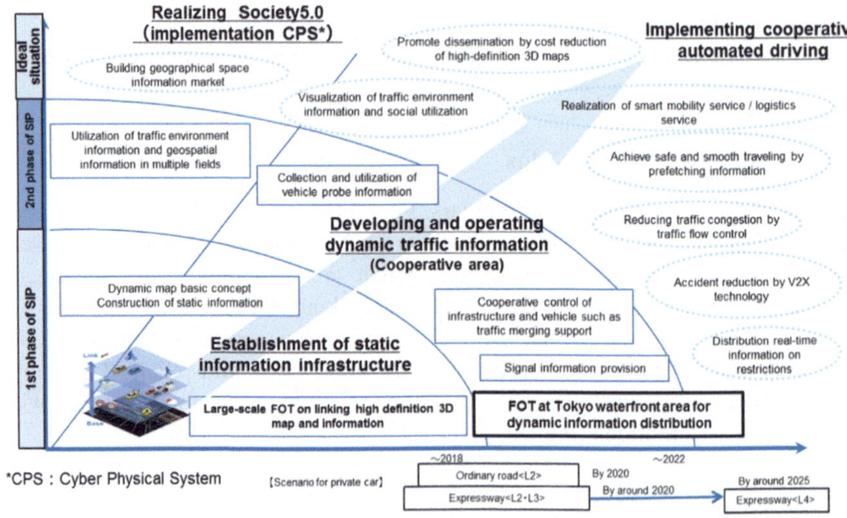

Fig. 9. Traffic environmental information famework

4.3.2 FOTs (Tokyo Waterfront City–Haneda Area)

FOTs will start in autumn 2019 in the Tokyo waterfront city area (general roads and Metropolitan Expressway in the Tokyo Waterfront City area/Haneda area) in cooperation with Japan Automobile Manufacturers Association. SIP-adus aims to increase public acceptance by involving local government, the general public, etc. (Fig. 10).

In Tokyo water front city:
Signal information from about 30 traffic lights is provided by DSRC. Vehicles are allowed to pass through intersections safely and smoothly based on the signal display and change timing information even in environments where recognition is difficult using in-vehicle cameras.

On the highway:
Providing vehicle information on the main lane from the road side for merging assistance. The speed and timing to enter the main lane are automatically adjusted to ensure safe merging.

In Haneda area:
The next generation Advanced Rapid Transit will be implemented on public roads by using automated driving technology in mixed traffic flow.

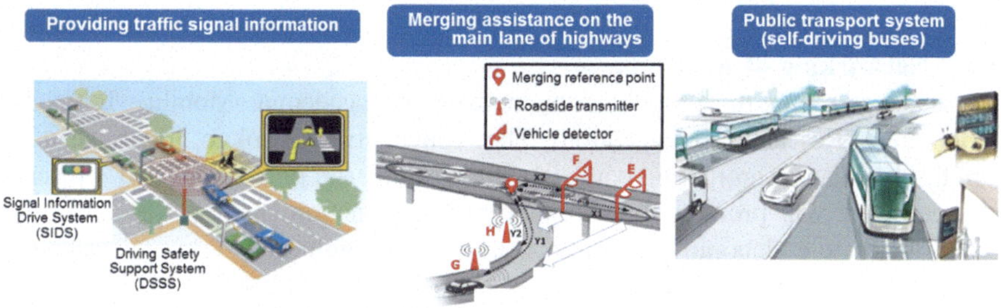

Fig. 10. FOTs (Tokyo waterfront city–Haneda area)

4.3.3 FOTs (Local Transportation)

In Japan, the progress of depopulation and aging in rural areas is a serious social issue. In rural area, SIP-adus plans FOT using simple and inexpensive automated vehicle for evaluation of social acceptance.

Long-term FOTs will be implemented in underpopulated areas, local communities, etc. through collaboration with businesses and local government to validate the effectiveness and business feasibility of automated driving in terms of logistics and mobility services (Fig. 11).

Fig. 11. FOTs (local transportation)

4.3.4 Virtual Environment for Safety Evaluation

SIP-adus will start the development of safety assessment methodology with JAMA in the second phase. Simulation tools for assessing the safety performance of automated driving in various traffic environments will be established. It is necessary for safety assessments to mix actual long-term driving tests and simulation.

4.3.5 Networking of Academia

SIP-adus is working to strengthen the network of academia. Mobility Innovation Collaborative Research Organization, The University of Tokyo, established the 'Mobility innovation promoting council' with 17 domestic universities and 3 research institutes. They will promote collaboration and information sharing. SIP-adus asked them to facilitate collaborative research with overseas entities.

5 Summary

SIP-adus is a five-year research program on connected and automated driving led by the Japanese government that first begun in 2014. Among relevant technical issues, cooperative field technologies were selected as the research themes of the first phase of SIP-adus. In the second phase that started in 2018, it expanded its operational domain from highways to public roads. Also, regarding logistic and mobility services, it is critical to implement automated driving service for solving social issues like mobility in rural area and shortage of truck and bus drivers. Field Operational Test are planned at Tokyo waterfront area for 2020. The network of academia will be strengthened and facilitate collaborative research with overseas entities.

References

1. Council for Science Technology and Innovation, Brochure, Cabinet Office of Japan (2017)
2. Pioneering the Future: Japanese Science, Technology and Innovation, Brochure, Cabinet Office of Japan (2018)
3. Sugimoto, Y., Kuzumaki, S.: SIP-adus: an update on japanese initiatives for automated driving. In: Meyer, G., Beiker, S. (eds.) Road Vehicle Automation 5. Springer, Cham (2018)

Developing End-to-End Regulation for Automated Vehicles in Australia

Marcus Burke(✉)

Future Technologies National Transport Commission,
National Transport Commission, Level 3, 600 Bourke Street,
Melbourne 3000, Australia
mburke@ntc.gov.au

Abstract. Australia's laws do not currently support the use of automated vehicles. Australia is currently working towards a goal of developing end-to-end regulation to support the safe commercial deployment of automated vehicles at all levels of automation. This chapter outlines the problem that Australia is trying to address and the key challenges. It outlines five key reforms that the National Transport Commission is leading on control, safety at first market entry, on road safety, insurance and data.

Keywords: Automated vehicles · Automated driving systems · Legislation · Safety · Australia · Policy

1 The Regulatory Challenge of Automated Vehicles

Automation offers potentially significant safety benefits in road transport. Driving is a risky activity – over 1200 people die each year on Australian roads [1].

Automated driving systems, that can carry out the entirety of the dynamic driving task for all or part of a trip, are not explicitly regulated in Australia. Australia has a range of legislation placing obligations on a human driver, which may create barriers to the commercial deployment of automated driving systems. A 2016 review found over 700 potential barriers to automated vehicles in current Australian federal, state and territory legislation [2]. Without reform, Australians may not be able to gain the potential safety, productivity and environmental benefits of automated driving.

There remain significant uncertainties for policy-makers about automated vehicles – when they will be ready for commercial deployment, what technology they will use, what business models will be developed and how the public will use this technology. Designing a legal framework with this level of uncertainty is extremely challenging. Governments will likely need to ensure that there is flexibility in regulatory frameworks to allow for different futures and track the deployment of the technology to adjust regulation if required.

G. Meyer and S. Beiker (Eds.): AVS 2019, LNMOB, pp. 23–29, 2020.
https://doi.org/10.1007/978-3-030-52840-9_3

1.1 The Australian Goal – End-to-End Regulation for Automated Driving

Australia is a federation, made up of six states, two territories along with the federal government. Federal, state and territory transport and infrastructure ministers meet as the Transport and Infrastructure Council ('the Council'). The Council agreed in November 2018 that Australia should aim to develop "End-to-end regulation to support the safe, commercial deployment and operation of automated vehicles at all levels of automation." [3] This goal makes safety the priority but recognizes that we need to develop a framework that:

- addresses all key legislative issues (end-to-end),
- supports not just trials, but commercial deployments and
- is general enough to cover all levels of automation.

A national goal ensures that all levels of government are working together to develop a nationally consistent framework. The National Transport Commission has led the policy development towards this goal and makes recommendations to the Council on key policy issues.

1.2 Rethinking How (and Whom) We Regulate

Current road transport law regulates vehicles separately from drivers. Current vehicle regulation governs the first supply of vehicles to the Australian market, largely aimed at manufacturers. In addition, in-service requirements, focused on vehicle owners and operators seek to ensure vehicles are maintained in a safe condition. Regulation of drivers also has a 'first supply' approach (driver licensing) and in-service elements (such as Road Rules) to ensure that vehicles a driven safely. These concepts are illustrated in simplified form in Fig. 1.

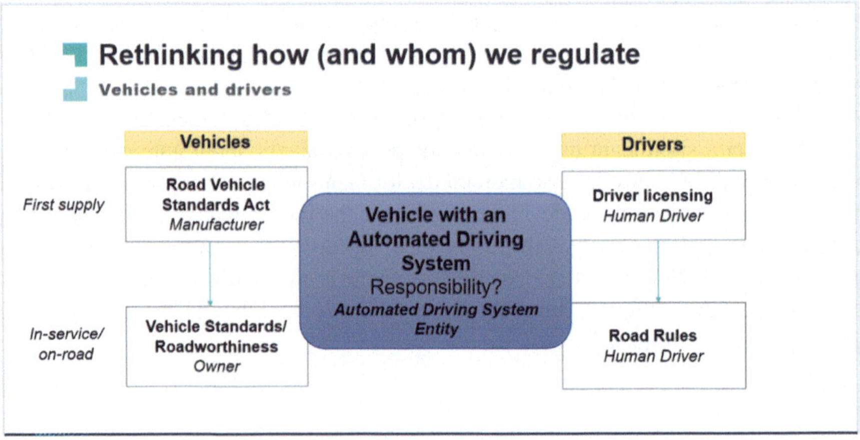

Fig. 1. Simplified regulation of vehicles and drivers and the areas an automated driving system would need to comply with.

The introduction of vehicles with an automated driving system that can carry out the driving task blurs these distinctions between vehicles and drivers in regulation. It also requires reconsideration of the responsible parties. The Council has agreed the need to create a new party, the Automated Driving System Entity (ADSE), as having responsibility for the automated driving system [4].

1.3 A New Responsible Party – Who Is the Automated Driving System Entity?

The Automated Driving System Entity, or ADSE, would be the company bringing the automated driving system to the Australian market and assuring its safety. This could be a traditional automotive manufacturer or a new technology company. This approach provides flexibility as the role is effectively self-selected - the legal entity that certifies that the Automated Driving System can safely perform the driving task in place of a human driver in Australia takes on the responsibility for the system. The ADSE would self-nominate by seeking approval for the ADS under the Road Vehicle Standards Act 2018 (Cth).

1.4 Five Key Automated Vehicle Reforms

Achieving the goal of end-to-end regulation for automated vehicles, requires reforms to a range of laws. Based on that goal, the National Transport Commission is working on five key reforms. These reforms seek to answer the following questions:

1. Who is in control of an automated vehicle when it is operating in automated mode?
2. How do we ensure automated vehicles are safe when they first enter the market?
3. How do we ensure automated vehicles operate safely throughout their life on the road?
4. How do we manage motor accident injury insurance for automated vehicles?
5. How do we manage access to data generated by automated vehicles?

Whilst the detail of legislation will be different in different countries, all jurisdictions will likely need to answer these questions. I will set out the current work to address each of these questions below. Subsequent reforms will be required as governments work through the detail.

2 Australia's Key Automated Vehicle Reforms

To achieve Australia's goal of end-to-end regulation for automated vehicles, we need to answer all five questions. We must also recognise that these questions are interlinked and will also connect to broader government policy areas. These include infrastructure, planning and pricing and changes required to criminal laws, passenger transport, freight and taxi legislation.

2.1 Who Is in Control?

An automated vehicle is a vehicle where the automated driving system is carrying out the dynamic driving task for all or part of the trip. Human drivers currently have a range of legal obligations in relation to the dynamic driving task, including following speed limits, stopping at red lights and avoiding pedestrians. Who should have responsibility for these legal obligations, when there is no longer a human driver?

The Transport and Infrastructure Council agreed in May 2018 that the ADSE, as the company responsible for the technology, should be considered in control of the vehicle when it is operating in automated mode [4]. As a result, the ADSE would have responsibility for the dynamic driving task obligations. This would be the case at SAE Levels 3, 4 and 5 [5]. At Level 3 (conditional automation) if control was handed back to the human driver, that driver would then be considered in control and have responsibility for those driving obligations. Figure 2 below illustrates this approach. This approach provides clarity as to which party is in control at any particular stage. However, there will be challenges assigning responsibility at the point of hand over of control between the human driver and the automated driving system.

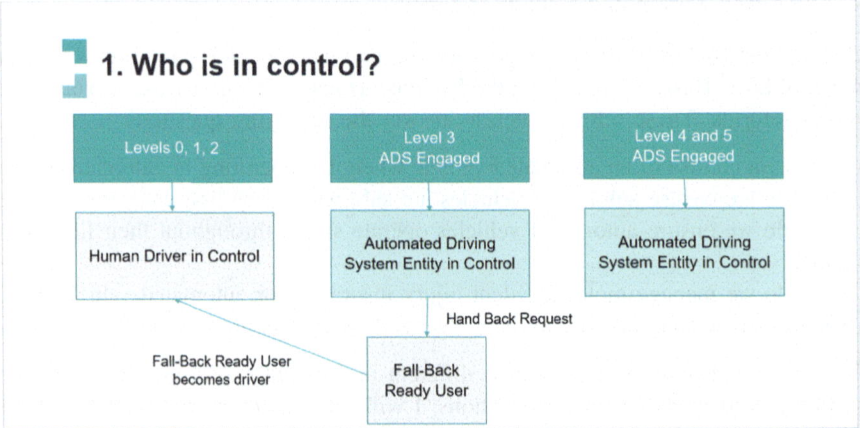

Fig. 2. Who is in control of an automated vehicle?

2.2 Safety at First Supply – How Do We Ensure That Automated Vehicles Are Safe When They First Enter the Australian Market?

New vehicles must meet minimum safety standards to enter the Australian market. Should there be safety standards for automated driving systems and if so, what should they cover?

In November 2018, the Council agreed to a policy for assessing the safety of automated vehicles at first supply [6]. Australia was one of the first countries in the world to agree such a policy [7]. The policy is a self-certification approach. Companies looking to bring automated driving systems to market in Australia will need to provide evidence against 11 safety criteria and meet three additional obligations [6]. The safety criteria are based in part on the United States Automated Vehicle Policy [8] along with research and consultation conducted in Australia. The 11 criteria are:

1. Safe system design and validation processes
2. Operational design domain
3. Human-machine interface
4. Compliance with relevant road traffic laws
5. Interaction with enforcement and other emergency services
6. Minimal risk condition
7. On-road behavioural competency
8. Installation of system upgrades
9. Verifying for the Australian road environment
10. Cybersecurity
11. Education and training

In addition, ADSEs will need to meet requirements on:

1. Financial solvency
2. Corporate presence
3. Data.

The approach provides flexibility for industry – it does not mandate a particular technology, level of automation, operational design domain or business model. The requirements provide confidence to Australian governments and the Australian public of the safety of these systems and the ability of the ADSE to support them. Finally, the approach provides flexibility to incorporate international standards as they are developed.

2.3 In-Service Safety for Automated Vehicles – How Do We Ensure Automated Vehicles Operate Safely Throughout Their Life on the Road?

Regulation at first supply can help ensure that only safe vehicles enter the market. However, that still leaves the question of how to ensure that automated vehicles continue to drive safely throughout their life on the road. How do you ensure that an automated driving system continues to operate safely five, ten or fifteen years after it enters the market?

There is a risk that automated vehicles will introduce new safety risks. In a federation like Australia there is also a risk of inconsistent state regulation, which could become a market barrier [9].

In 2019, the National Transport Commission conducted a public consultation on which parties influence on-road safety, how those parties were covered by existing laws, what if any were the gaps in regulation, and who would be the regulator for automated driving systems. A key issue was whether Australia's laws should include a general safety duty for ADSEs to ensure the safety of their automated driving systems. The National Transport Commission will provide recommendations to the Council on these issues in mid-2020.

2.4 Motor Accident Injury Insurance and Automated Vehicles

Automated vehicles are expected to be much safer than vehicles today. But there will still be crashes. This raises the issue of how a crash involving a vehicle with an automated driving system would be covered by insurance, in particular when the crash results in injury or death.

All Australian states and territories have compulsory third party insurance to cover injuries caused in motor vehicle crashes. There are significant differences between these systems. Some are fully public, some are private; some are fault-based, some are no-fault. Each has different definitions and thresholds.

The NTC publicly consulted on motor accident injury insurance and automated vehicles in 2018. In August 2019, the Council agreed to a national approach to motor accident injury insurance and automated vehicles, based on the principle that no person should be worse off, financially or procedurally, if they are injured by a vehicle whose automated driving system was engaged, than if they were injured by a vehicle controlled by a human driver [10]. The application of this principle would mean that existing schemes would cover crashes involving an automated driving system, although further work is required to ensure that insurance schemes can recover claims from appropriate parties. At the time of writing, these recommendations were still to be considered by responsible ministers for the schemes in some states and territories.

2.5 Access to Automated Vehicle Data

Automated vehicles will generate greater volumes of data and more detailed data than conventional vehicles today. This potentially creates opportunities for new services and better ways to run transport systems. It also raises potential privacy challenges. The NTC has examined the potential privacy implications of connected and automated vehicles and proposed principles to guide future reform [11]. The NTC is also carrying out further work on the opportunities created by in-vehicle data.

3 Conclusion

This chapter has outline five major reforms to create a legal framework for the safe commercial deployment of automated vehicles. The reforms, if completed carefully and implemented appropriately, can provide a flexible system that creates certainty for industry and the public and supports both safety and innovation. All countries will need to examine a similar set of questions within the context of their own legal systems and transport policies.

Beyond these five reforms, significant further work will be required, to amend driving offences in criminal laws; ensure that passenger transport legislation can support automated vehicles and to ensure that legislation supports automated freight.

The end-to-end regulatory framework should aim to:

1. Ensure that safety is the priority
2. Provide legal certainty
3. Ensure responsibilities are placed on parties best able to manage the risk

4. Be outcomes-based, rather than prescriptive
5. Be internationally aligned
6. Be technology neutral, as well as business model and application neutral
7. Provide the ability to evolve over time as the technology and businesses evolve.

National policy agencies will need to liaise with their international counterparts to share knowledge and best practice on common issues. Governments will also need to ensure ongoing discussions with industry, to make sure that regulation is fit-for-purpose and supports, rather than hinders the deployment of safety technology.

The opportunity for reducing the 1200 deaths each year on Australia roads is significant. It is important that we build the regulatory framework that ensures that Australians can gain the safety, productivity and environmental benefits of this technology when it is ready for commercial deployment.

References

1. Australian Bureau of Infrastructure, Transport, Cities and Regional Development: Road Deaths in Australia Monthly Bulletin, December 2019
2. National Transport Commission: Regulatory reforms for automated road vehicles, November 2016
3. National Transport Commission: Towards 2020: transport ministers approve vital next phase of automated vehicle regulation. Media Release, 13 November 2017
4. National Transport Commission: Changing driving laws to support automated vehicles. Policy Paper, May 2018
5. Society of Automotive Engineers: Taxonomy and Definitions for Terms Related to Driving Automation Systems for On-Road Motor Vehicles (2018)
6. National Transport Commission: Safety Assurance for Automated Driving Systems. Decisions Regulation Impact Statement, November 2019
7. KPMG: Automated Vehicle Readiness Index (2019)
8. United States Department of Transportation: Preparing for the Future of Transportation: Automated Vehicles 3.0, 2018. This has now been superseded by United States Department of Transportation, Ensuring American Leadership in Automated Vehicle Technologies: Automated Vehicles 4.0, January 2020
9. National Transport Commission: Consultation Regulation Impact Statement: in-service safety for automated vehicles, July 2019
10. National Transport Commission: Motor accident injury insurance and automated vehicles policy paper, August 2019
11. National Transport Commission: Regulating government access to C-ITS and automated vehicle data, August 2019

Part II Business Models and Operations

Business Models for Shared and Autonomous Mobility

William Riggs[1] and Sven A. Beiker[2(✉)]

[1] University of San Francisco, 2130 Fulton Street, San Francisco, USA
wriggs@usfca.edu
[2] Stanford Graduate School of Business, Stanford University, 1102 Clark Way, Palo Alto, USA
beiker@stanford.edu

Abstract. Shared autonomous systems are an opportunity for cites to improve mobility, yet little discussion has framed their business model and how public and private stakeholders can plan for their deployment. This chapter frames the range of shared services as well as the anticipated opportunities and challenges to shared autonomous mobility ecosystems. It anticipates opportunities of "platform" and opportunities of "place", and frames these within the context of autonomous vehicle development, concluding that partnerships, both business-to-business and across sectors, are a key to solving many business challenges for shared and autonomous mobility.

Keywords: Shared mobility · Autonomous vehicles · Business models · Economics · Market trends · Automobiles

1 Introduction

Private car owner ship is stagnating or even in decline in some key markets, especially in metro areas of developed economies such as the United States, Europe, or Japan [1]. This trend is driven by several components, including economic reasons, changing consumer preferences, and regulation [2–9]. This trend appears to be especially prevalent among younger consumers as the decline in U.S. license holders among 16–24-year-old consumers shows [10, 11]. However, it is subject to debate if these are short-term developments, more or less linked to economic cycles such as – in particular and most recently – the 2008 recession, or if these are early indicators of a profound long-term change in mobility preferences and ownership structures. A third hypothesis might be, in the absence of long-term data, that sharing will mostly become an additional component of the mobility portfolio without changing the traditional ownership and usage model at large. In other words, it is unclear if ownership of a private automobile is "a thing of the past" or if sharing is simply complementing it.

Meanwhile, shared mobility options appear to be booming in many key global markets [12, 13]. The shared mobility spectrum is ranging from car sharing and ride sharing, to ride hailing and micro transit, and all the way to micro mobility. It is at times difficult to stay abreast of developments with new offers coming to the market

G. Meyer and S. Beiker (Eds.): AVS 2019, LNMOB, pp. 33–48, 2020.
https://doi.org/10.1007/978-3-030-52840-9_4

and others consolidating or going out of business at the same time. Table 1 gives an overview of the shared mobility spectrum with primary categories that are commonly used to structure the field.

Table 1. Categories of shared mobility offers and examples of current offerings on the market, adapted from [14].

Category	Description	Examples
Car sharing, station based	Initial type of car sharing where vehicles are picked up and returned at the same location; typically through an hourly rental	City Car Share, Hertz 24/7, Zipcar
Car sharing, free floating	2nd generation car sharing; vehicles can be picked up and dropped off in different locations (zones); paid by the minute	Enjoy, GIG, ShareNow
Car sharing, peer-to-peer	Crowd-based car sharing where individuals can rent out their individual vehicle to others at their discretion	Getaround, Turo
Ride hailing	Platform where individuals can hail and pay for a ride from a professional driver or "gig-worker" through an app	FreeNow, Grab, Gett, Lyft, Uber, Didi Chuxing
Ride sharing	Extension of ride-hailing where individuals can be matched in real-time to share rides with others going on similar route	BlaBla Car, Curb, Lyft Line, Scoop, Uber Pool, Via
Micro transit	App-based shuttle services, typically in a van-size vehicle; some with dynamic routing, others with semi-fixed routes	MOIA, Chariot (closed 2019), Via
Micro mobility	Bicycles and scooter sharing, some of those electric; station-based or floating, paid by the minute/trip/subscription	Bird, Jump, Lime, …

As shared mobility offerings are expanding into the market, two other key aspects should be considered: (1) shared mobility offerings struggle with profitability and are therefore constantly innovating their business models to improve economics; and (2) automation of automobiles with the goal of self-driving cars has much potential to completely change mobility patterns. The combination of those two trends is particularly interesting because many shared mobility providers are betting that automation be the most promising solution to bring profitability to their business model [14–16]. Looking collectively at those two trends, combining sharing and autonomy, a discourse seems to be in order to discuss how the business models for shared and autonomous mobility have the potential to redefine use cases and revenue streams as we are entering the new mobility paradigm.

Table 2. Levels of vehicle automation and the status of public operations [17, 18]

Automation level	Description	Status
0	**No driving automation** Vehicle control: human Operation monitoring: human Fallback for error: human Situation/time limitation of system: N/A	**On the market** (Any standard vehicle, esp. w/out advanced control system)
1	**Driver assistance** Vehicle control: human and system Operation monitoring: human Fallback for error: human Situation/time limitation of system: yes	**On the market** (e.g. adaptive cruise control <u>or</u> lane centering)
2	**Partial driving automation** Vehicle control: system Operation monitoring: human Fallback for error: human Situation/time limitation of system: yes	**On the market** (e.g. adaptive cruise control <u>and</u> lane centering)
3	**Conditional driving automation** Vehicle control: system Operation monitoring: system Fallback for error: human Situation/time limitation of system: yes	**Current frontier** (e.g. introduced in very limited situations such as parking, stop-and-go)
4	**High driving automation** Vehicle control: system Operation monitoring: system Fallback for error: system Situation/time limitation of system: yes	**Pilot testing** (e.g. tests of ride-hailing services/autonomous vehicle companies)
5	**Full driving automation** Vehicle control: system Operation monitoring: system Fallback for error: system Situation/time limitation of system: no	**No prediction** (Unlimited operation characteristic makes forecasts impossible today)

The examples in Table 2 show how far the industry has progressed with automated driving and what can be expected soon to impact the business model of shared mobility. This chapter aims to provide context to what extent automation can actually have an impact on the financials of shared mobility in which providers are seeking a sustainable business model.

Finally, with the promise upon us that shared and autonomous mobility will fundamentally change uses cases and revenue streams, one also needs to consider broader implications when personal mobility becomes ultimately safe, affordable, and convenient. It can only be imagined that passenger miles – and quite possibly vehicle miles as well – will increase drastically, which basically means more traffic, more congestion, and more environmental impact. Such rebound effects were observed initially as the Jevon's Paradox following efficiency gains in coal-powered engines and later also in other fields, i.e. that consumption increases when efficiency increases [19]. Therefore, stakeholders need to consider the implications and sustainability of those business models—including both, economic and environmental sustainability.

This chapter dialogues trends and explores if gains in automation can lead to more sustainable business models. These are discussed within the context of challenges and opportunities for urban planning, the industry landscape, societies, and more. Potential models are then framed within the context of "platforms" and "place"—ultimately leading to potential suggestions that might turn the losses of shared mobility providers into new operational savings and more fiscally responsible revenue streams.

In order to assess this potential, one also needs to consider the current and near-future status of automated driving, especially in context of personal mobility solutions. Table 2 gives an overview of examples in automated driving that are indicative of where the industry is at and what can be expected soon to impact the business model of shared mobility. In the end the question is if automation can be implemented soon enough to actually turn the losses of shared mobility providers into profits or if costs can be driven to a point where private auto-makers can sell vehicles to private owners that can be shared with other riders when not in use.

2 Challenges of Shared and Autonomous Mobility Today

As a precursor automated vehicles, shared services today have changed the ways that people get around cities [8] These new ecosystems have combined traditional modes like public transportation with digital technology, resulting in digital "ridesourcing" services, also known as Transportation Network Companies ("TNCs") [12, 20, 21]. As many of these services mimic automated services [22] they provide important insights when considering the business model for automation.

2.1 Challenges of Increases in Trips and Vehicle Miles Traveled

First, research indicates that shared services have been connected with increases in trips, at least partially connected to decreases in cost. In 2017 Schaller suggested these services generated an additional 600 million vehicle miles from 2013 to 2016 in New York, most of which would not have existed otherwise [23]. He found as many as 61% of ride-hailing trips would not have been made at all, or would have otherwise been made via walking, biking, or transit [24].

This has been confirmed by other research in Boston [25] and elsewhere [26]—where contrary to traditional narratives that Uber or Lyft may be stealing rides from taxis, most trips actually come from transit, walking, cycling or did not exist at all. This increase in use from ridesharing/ridesourcing services is complicated for many planning and transportation professions—particularly in the arena of emissions and sustainability since an increase in trips and VMT implies an increase in use.

2.2 Challenges and Questions of Access

Likewise, cities, urban planners, and businesses continue to wrestle with the last mile issue facing populations that reside in areas that are dependent on public transit systems. Multimodal transportation along with vehicle-to-infrastructure considerations can improve urban congestion and contribute to transit efficiency. As a first mover in ridesharing, Uber has filled gaps in public transit that has transformed travel and travel behavior [27–29]. This work focuses on this last mile issue faced by areas in the U.S. that are lower in population densities than cities and require mode options to travel between transit hubs and homes.

Long distance pick-up fees may be getting subsidized by public private partnerships like between New Jersey and Uber for those shuttling to and from commuter stations [30], but the fees affect the travelers seeking a ride to areas other than transit hubs where parking is a struggle. With drivers more spread out, suburban riders pay a premium that does not make UberPool or UberX viable mode options to replace driving and, thus, emissions. These policies might incentivize drivers to leave cities for the suburbs—yet it remains unclear if the demand strong enough to drive the supply.

In addition to sparsely populated, we would also like to focus on disadvantaged communities. Are TNC trips helping serve distal, underserved areas across markets. Research in Boston and LA suggests this [31, 32] and that they are helping to serve traditionally disadvantaged communities. But it is also unclear how TNC trips are serving suburbia and what types of land usage exist at the origin and the destination of trips. Research shows that mode substitution is happening from a variety of sources, including taxis, transit, biking and walking and that they are a complement to these services [33–35].

Table 3. Examples for autonomous driving programs targeting shared and "transit-like" mobility (Note: most use a disaggregated network)

Vehicle type	Description	Company
Transit bus	Traditional bus with fixed routes and schedules but without human driver, instead automation technology	Daimler, Volvo
Shuttle	Fixed route with potentially flexible schedule (on-demand), shuttles or buses	EasyMile, Navya, Optimus, VIA
(Robot-) Taxi	Flexible route and schedule, limited area (geo-fenced)	Lyft, May Mobility, nuTonomy, Uber, Waymo
Private car	TBD – Vehicles w/summon functionality; on-demand; flexible disaggregated network; public or private systems	Tesla

2.3 Challenges of Economics

Finally, one of the key challenges of shared mobility services is economic—the unit economics of serving many passengers on door-to-door trips has not been fully vetted. No shared mobility platform, from cars, to bikes and scooters, has demonstrated a pathway to a profitable business model [36]. There have been ample debates over wages and two sided markets for labor [37, 38], yet at the same time the cost of rides have continued to decline.

As automated vehicles enter shared fleets it remains to be seen if this economic tension is resolved particularly given the large amount of capital recovery that will be required to make autonomous vehicles profitable. The technology continues to need refinement, infrastructure and components remain cumbersome to scale and many liability and ethics questions remain. Some estimate that Waymo alone may be spending over $1 billion a year to develop their technology [39]—and with these pressures and the ever increasing demand for rides—profitable business models could be years in the making.

3 The Opportunities of Shared and Autonomous Mobility

3.1 Opportunities of Platform

Yet at the same time that costs are being incurred, there is ample opportunity for automation. In addition to increase accesses two of the largest benefits are increased safety and reduced costs. The vehicles will reduce the 37,000 fatalities that occur as a result of collisions each year just in the United States. AVs will also decrease the cost of mobility making more rides available, in more locations for less. This phenomenon may compound the increase in trips and VMT, and may be largely driven by the unique capabilities of hardware on specific vehicles that allows for more resolute detection, followed by software anticipation and then component actuation. While there is ample dialogue and debate about LIDAR, sonar, and photo-based detection, the greatest benefit in terms of latency may come from reduced delays in actuation—for example decreased delay in evasive turning or braking.

At the same time there is a parallel opportunity of electrification that promises that many of these new trips maybe carbonless, at least on a local level. Governments at all levels and the market have continued to underestimate consumer demand and express skepticism about electric vehicles [40]. Put simply, they are no longer a fad. In 2018 and early 2019 no one would have believed that Tesla would have exceeded all expectations in production goals and become the most valuable car company ever after delivering the Model 3 [41]. Now, many European companies, including Volvo and VW, are following and preparing to phase out fossil fuel vehicles while others like Ford and BMW are struggling to catch up. While these two companies may have some of the "Ten Electric Vehicles to Watch in 2020", they are still making up lost ground [42].

Likewise a great frustration for transit agencies in recent years is that they have gotten "beaten" by TNCs. They have been outcompeted by services based on better, more reliable, more convenient, door-to-door rides that can be accessed from the palm of our hand. Many transit champions want us all to enjoy the communal bliss of taking the bus [43] yet that is not a universally held value across society or demographic cohorts [34]. As such many transit agencies, who employ people who share those same values, have continued to resist rethinking platforms and services, and learning the lessons of TNCs and the changing demands of customers.

A handful of large and small transit operators have been finding that they can use large mass transit platforms in parallel with smaller door-to-door services and make the system more efficient, convenient and reliable. One of the largest examples is BVG in Berlin which has deployed a large-scale, app-based, on-demand, last-mile service that connects to existing rail and is beginning to use small self-driving buses [44]. Likewise, while debate continues on free and reduced fares [45], cities like Monrovia, CA (Go Monrovia) and Dublin, CA (Go Dublin) have been able to reduce costs by providing on-demand services and optimize trunk-line performance through partnerships with TNCs [46, 47]. All of this is happening as private sector companies continue to be frustrated with reliability and capacity limitations of existing transit and provide their own mass transit services. As Table 3 illustrates many companies are targeting this concept of a reinvented vision of disaggregated transit—something that could open up access to places in cities that previously had limited access and generated discussions about urban development to meet acute housing needs in many megaregions; in other words the density of the transit network increasing the opportunities for housing production [48, 49].

3.2 Opportunities of Place

In addition to these platform evolutions, new disruptive transportation will transform cities and streets spatially. Opportunities also exist to connect individuals to jobs and change the way cities organize space and optimize trips [50, 51]. According to the Three Revolutions report by UC Davis, without vehicle sharing we could see a 15% 20% increase in overall vehicle travel, assuming a 50% reduction in the personal cost of travel [52]. Even the optimistic vehicle sharing scenarios, which could lead to a 90% drop in vehicles overall, could still lead to a 10% increase in travel [53]. This yields an opportunity to rethink urban roadway allocation and redesign street for automation but also walking, biking and transit [9, 17, 54–56]. As suggest by Schlossberg et al, there could be opportunities to thin lanes, reduce parking and think about shared streets [57].

Vehicles might not only reduce the need for parking but also the size of space on roads occupied by vehicles [57, 58]. Wide city streets could be narrowed, adding more space for landscape elements or even new buildings in some circumstances [55, 56, 59]. In dense cities like Washington DC or San Francisco, right-sizing roadways could create additional space for residential or retail development and pedestrian activity, similar to Las Ramblas in Barcelona. In overbuilt cities such as Detroit, green infrastructure or multi-use paths could continue to take the place of some roadways and unused parcels of land.

4 Current Activities to Move to Shared and Autonomous Mobility

While Tables 1 and 2 provided already overviews on the shared mobility market and levels of automated driving, respectively, it is indicative how those two trends are already being merged in some early examples. Circle back to Table 3, one of the primary types of shared and autonomous vehicles are (1) automated transit vehicles such as buses and (2) autonomous shuttles with fixed routes. Further, there are (3) so called "robotaxis" that service a specific area free of determined routes or schedules and there are (4) summon functions to bring a shared vehicle to the user. It is important to note that these primary examples are all still in relatively early testing phases or at best in pilot trial mode. This in return means that none of them are offered to the general public as a means of transportation and are therefore, at this point not a readily available mobility solution. Yet, as they are getting deployed, there be more opportunities to harness private vehicles, two wheelers or other platforms for more public use; for example Tesla's roadmap to create a ridesharing network in parallel with Autopilot development [60].

In context of Table 3, the following four examples provide more detailed insights into what is already on the market or can be expected soon. Those examples serve as a good general overview because they are assumed to be of particular interest to the reader given that the companies behind them get much media, analyst, and industry attention. Therefore, one can maintain they are quite indicative of the status of the shared and autonomous mobility paradigm.

Back to Tesla as an example; known thus far mostly as a manufacturer of high-end and privately-owned electric vehicles, the company appears to be actively pursuing "robotaxis" [61]. They have publicly disclosed plans to offer technology enabling vehicle sharing and robotaxi operation as early as 2020. The goal is assumed to expand the value proposition of an automobile as a personal mobility solution to one that lets private owners earn income with their vehicles when not using it themselves or to enable shared fleets for commercial operators using a prefabricated platform. Both value expansions would address particular challenges related to mobility, which are for instance that personally owned vehicles remain unused more than 90% of the time [62] and that shared mobility is often accumulating losses, to a large extent due to the cost of human labor [15, 63].

Tesla is also notable for their "Summon" and "Smart Summon" functions [64], which lets the vehicle roll driverless to its owner or user so that he or she can wait at the entrance of a parking lot and the Tesla comes to him or her without direct human interaction. While this might appear to be a rather small step toward a truly shared and autonomous mobility paradigm with entirely new business propositions, it is to say that such summon function is a necessary step in this direction because it enables sharing vehicles among users as the vehicle can be summoned in a very convenient way. A direct revenue stream might be difficult to attach to this kind of automation and sharing, but the enabling character as the foundation for more functionality in the sense of the new mobility paradigm should not be discounted.

For mobility services such as Uber and Lyft in the United States the motivation to complement their existing shared mobility offerings like ride-hailing, ride-sharing, or others is to improve their economics once the cost of human labor (i.e. the cost of a driver) becomes negligible. It has been stated many times that the business model of low-cost ride-hailing, i.e. a service that is priced significantly below traditional taxi and going to compete with public transit, cannot be achieved with the traditional service paradigm of driver-reliant taxis or schedule-constrained busses or trams. But to get to a mobility paradigm that combines the flexible schedule of a taxi and the cost advantage of a public bus, cost has to be decreased. And with human labor making up around half the cost of ride-hailing [63], driverless mobility has become subject of intense research and development toward autonomous vehicles. Uber for instance pointed out in this context that the company might never be profitable unless autonomous vehicles can be used to improve the business case for shared mobility [16].

An additional example is Waymo, Google/Alphabet's self-driving automobile division. While Waymo's true motivation still remains largely subject to speculation, even after 10 years after its inception, the beforementioned value expansions for an automobile can assumed to be stimulus for the corporation as well. Other assumed benefits are the extension of the company's core business of data and advertising to previously untapped market segments. In that sense, providing mobility to users at an unprecedented cost can be expected to generate additional revenue streams of user data, advertising revenue, and service consummation.

5 Partnerships to Shape Shared and Autonomous Mobility

There are many more examples to study regarding the integration of automation and sharing to usher in a new mobility paradigm. Some of the well-covered examples include (in alphabetical order): Lyft, May Mobility, Optimus, Zoox. In studying those respective ventures, one will find again and again very similar drivers to pursue shared and autonomous mobility, which are cost and convenience, yet they also harness a common theme of partnership.

As alluded to previously there are many partnerships that can accelerate and promote automation—both capitalizing on the product side of the vehicle and the current activities to accelerate levels of automation, as well as the place-related side. We already see many industry partnership to accelerate software development for automation (for example Waymo and Chrysler, Daimler and Bosch, Continental and EasyMile, Daimler and BMW) but expect this to continue to extend both the platform (physical and virtual) as well as the place and for it to cross public and private sectors.

In the realm of platforms, the increasing convergence of electrification and automation is likely to accelerate with many cities incentivizing level-2 charging and relaxing permit fees and requirements for electric vehicle infrastructure. As this trend continues, the sharing economy may likely include a charging component for charging with potential offerings of on or off-street public and private parking.

Similarly, with regard to place, we are already seeing partnerships between cities and autonomous vehicle developments (e.g. Waymo with Chandler/Phoenix, nuTonomy with Boston) and this is likely to continue, along with more localized autonomous regulation initiatives that account for local controls over driving behaviors. These partnerships are particularly important as it becomes clearer that increasing vehicle occupancy is important for high-volume, near free shared autonomous mobility—and in that light transit operators must be open to tap into these trends to learn-from/partner-with new mobility [8]. This is the vision that has been cast in early 2020 announcements from Cruise Automation with the Cruise Origin—an efficient electric autonomous vehicle built for on-demand, transit-like service [65].

Transit operators should explore partnerships with TNCs who can help with system efficiency and service reliability on central lines. They should rethink buses in terms of cost-effectiveness and environmental impact and consider lightweight or tactical solutions where running trunk-line service is not efficient. They should think of transit service in a comprehensive way and explore partnerships with private/corporate transportation providers, leveraging systems for the benefit of all. They should establish comprehensive and creative Transportation Demand Management (TDM) and travel behavior programs [66] that experiment with both financial as well as social incentives —which can be equally if not more powerful than monetary norms.

6 Timelines, Conclusions and Future Outlook

Given these opportunities, activities and potential partnerships where does this lead planners, engineers, policy makers and citizens? What might a future look like when these respective offerings and new markets emerge, shaping new business models for mobility? As has been pointed out throughout this paper, shared mobility solutions have already become common in many forms and markets (car sharing, ride sharing, ride hailing, micro mobility) giving consumers the possibility to be mobile without owning a vehicle. Yet there are still developments in "platform" and in "place" and partnerships to be made that will shape this future.

It should be emphasized, however, while autonomous driving or "self-driving" technology has, in fact, been demonstrated in many pilot programs, there still remains uncertainty. At this point most solutions are not available for general public use but only to registered users at best. While, Waymo is the assumed frontrunner in this field, its Waymo One program in Arizona remains limited to certain users, routes, times [67]. And as we have argued, the business model and timeline for deploying this technology need refinement. Cruise and others are working across OEMs to build new mobility-as-a-service platforms like Origin [68].

To undertake forecasts for shared and autonomous mobility, one needs to consider the general framework of automation levels in Table 2, keeping in mind that only Level 4 or 5 automation have the potential to actually replace the human driver and that the most savings in operating costs are achieved in that state. To provide a better understanding of the deployment roadmap, Fig. 1 shows projections by well-known technology and market analysts for the new vehicle sales share of Level 4 and 5 automated driving.

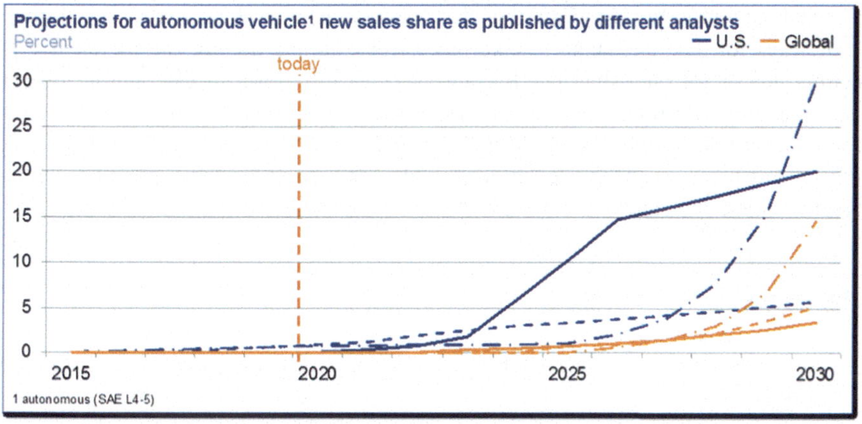

Fig. 1. Forecasts for automated driving Level 4–5 by different institutions, selected and aligned by silicon valley mobility

Figure 1 illustrates that the forecasts for market penetration of Level 4 and 5 automated driving differ substantially between analysts and regions. Hence it is unclear when automation technology will be readily available to bring the expected improvement to the business model of shared mobility with the new mobility paradigm ushered in. This uncertainty for Level 4 and 5 forecasts is not surprising given the key point in Table 2—that Level 2 automation is well established in the market and Level 3 has just been introduced in some very limited situations [69]. And still, Fig. 1 also shows that Level 4, which was originally expected for the early 2020s [70–72], might now gain substantial market share around the mid-2020s according to more recent announcements [73–76].

The reasons for this delay relate to our argument that "platform" refinements are important to the savings that will help achieve sustainable business models. Many of the delays of automated driving deployments are challenges in technology solutions (e.g. sensors, artificial intelligence), infrastructure installations (e.g. communication standards, zoning for operation), regulation (e.g. certification, permitting), and others that only surface when moving from early demonstrations of the technology to current pilot programs. Dmitri Dolgov, CTO of Waymo summarized this point concisely when he said that, "the challenge of actually building a real product and deploying it so that people can use it has turned out to be more difficult than I expected" [77].

These challenges and forecasts offer an opportunity for our dialogues place-based and partnership strategies to be the focal point of near-term shared mobility business models—these models then being improved as autonomous technology is refined and operational efficiency gained in the second half of the 2020s. In this aspect it remains clear that shared mobility must find new revenue streams and business opportunities to improve profitability in the near term as highly automated vehicles develop—much in the same way that search engine operators of the early 2000s experienced a step change in their economics through a new paradigm when they ventured into online advertising.

For the new mobility paradigm this could involve new products to delivery during a ride or ways for data exchange to take place (for example trading rides exchange for watching ads or for data sharing). It could also take the form of discounting rides to certain consumer destinations (e.g. free rides to department stores or restaurants with the purchase of an item or meal). At the end of the day however it is clear that for the platform and place, both, the technology and a change in business practice, will be needed to shape the future of a sustainable business mode for shared mobility.

References

1. Naughton, K., Welch, D.: This Is What Peak Car Looks Like. Bloomberg.com. Accessed 28 Feb 2019
2. Reuters, T.: Ditching personal cars for ride sharing. Thompson Reuters (2017). http://fingfx.thomsonreuters.com/gfx/rngs/AUTOS-RIDESERVICES/0100419J2RP/index.html. Accessed 01 Feb 2020
3. Grosse-Ophoff, A., Hausler, S., Heineke, K., Möller, T.: How Shared Mobility will Change the Automotive Industry. Mckinsey & Company, New York (2017)
4. Garfield, L.: Cities are going car-free around the world - business insider. Business Insider (2017). https://www.businessinsider.com/cities-going-car-free-2017-2. Accessed 01 Feb 2020
5. Jaffe, E.: 6 big European cities with plans to go car-free. CityLab (2015). http://www.citylab.com/cityfixer/2015/10/6-european-cities-with-plans-to-go-car-free/411439/. Accessed 02 Oct 2017
6. O'Sullivan, F.: Stockholm starts a friendly rivalry over car-free planning. CityLab (2017). https://www.citylab.com/design/2017/05/stockholm-pedestrian-downtown-plans-oslo/526464/. Accessed 27 June 2018
7. Goodyear, S.: In stockholm, vision zero 2.0 is a dream for a car-free future. CityLab, 24 March 2015. http://www.citylab.com/commute/2015/03/how-stockholm-became-the-ultimate-walkable-city/388433/. Accessed 23 July 2015
8. Riggs, W.: Disruptive Transport: Driverless Cars, Transport Innovation and the Sustainable City of Tomorrow. Routledge, London (2019)
9. Riggs, W., Appleyard, B., Johnson, M.: A design framework for livable streets in the era of autonomous vehicles. In: Proceedings of the 98th Annual Meeting of Transportation Research Board, Washington, D.C. (2019)
10. Roberts, A.: Driving? The kids are so over it. Wall Str. J. (2019). https://www.wsj.com/articles/driving-the-kids-are-so-over-it-11555732810
11. Henderson, T.: Why many teens don't want to get a driver's license. PBS NewsHour, 06 March 2017. https://www.pbs.org/newshour/nation/many-teens-dont-want-get-drivers-license. Accessed 01 Feb 2020
12. Shaheen, S.: Shared mobility: the potential of ridehailing and pooling. In: Three Revolutions, pp. 55–76. Springer (2018)
13. Yakovlev, A., Otto, P.: The Future of Mobility - Shared Mobility. Ipsos, Paris (2018)
14. U.S. Federal Transit Administration: Shared mobility definitions. Federal Transit Administration, 07 December 2016. https://www.transit.dot.gov/regulations-and-guidance/shared-mobility-definitions. Accessed 01 Feb 2020
15. CB Insights: How Uber makes—and loses—money. CB Insights Research (2018). https://www.cbinsights.com/research/report/how-uber-makes-money/. Accessed 01 Feb 2020

16. Uber: Form S-1, Uber Technology. US Securities and Exchange Commission (2019). https://www.sec.gov/Archives/edgar/data/1543151/000119312519103850/d647752ds1.htm. Accessed 01 Feb 2020
17. Crute, J., Riggs, W., Chapin, T., Stevens, L.: Planning for autonomous mobility. American Planning Association, Washington D.C., PAS 592 (2018)
18. SAE: Taxonomy and Definitions for Terms Related to Driving Automation Systems for On-Road Motor Vehicles. SAE International, Warrendale (2018)
19. Alcott, B.: Jevons' paradox. Ecol. Econ. **54**(1), 9–21 (2005)
20. Shaheen, S., Chan, N., Rayle, L.: Ridesourcing's impact and role in urban transportation. Access Mag. (2017). https://www.accessmagazine.org/spring-2017/ridesourcings-impact-and-role-in-urban-transportation/
21. Cohen, A., Shaheen, S.: Planning for shared mobility. American Planning Association, Planning Advisory Services, vol. 583 (2017)
22. Harb, M., Xiao, Y., Circella, G., Mokhtarian, P., Walker, J.: Projecting travelers into a world of self-driving cars: naturalistic experiment for travel behavior implications. In: Proceedings of the 97th Transportation Research Board, Washington D.C. (2018)
23. Schaller, B.: Unsustainable? The growth of app-based ride services and traffic, travel and the future of New York City. Rep. Schaller Consult. Brooklyn (2017)
24. Clewlow, R.R., Mishra, G.S.: Disruptive transportation: the adoption, utilization, and impacts of ride-hailing in the United States. University of California, Davis, Institute of Transportation Studies, Davis, CA, Research Report UCD-ITS-RR-17-07 (2017)
25. Gehrke, S.R., Felix, A., Reardon, T.G.: Substitution of ride-hailing services for more sustainable travel options in the greater Boston region. Transp. Res. Rec. (2019). https://doi.org/10.1177/0361198118821903
26. Hampshire, R., Simek, C., Fabusuyi, T., Di, X., Chen, X.: Measuring the impact of an unanticipated disruption of Uber/Lyft in Austin, TX. Social Science Research Network, Rochester, NY, SSRN Scholarly Paper ID 2977969 (May 2017)
27. Lazo: Uber, Lyft partner with transportation authority to offer paratransit customers service in Boston. Washington Post (2016)
28. McMahon, J.: 5 ways city transit agencies have exploited Uber And Lyft. Forbes (2018). https://www.forbes.com/sites/jeffmcmahon/2018/09/06/5-ways-city-transit-agencies-have-found-to-exploit-uber-and-lyft/. Accessed 14 Dec 2018
29. Hall, J.D., Palsson, C., Price, J.: Is Uber a substitute or complement for public transit? J. Urban Econ. **108**, 36–50 (2018). https://doi.org/10.1016/j.jue.2018.09.003
30. Hawkins: Coming soon to the Uber app: bikes, rental cars, and public transportation. The Verge (2018). https://www.theverge.com/2018/4/11/17220408/uber-jump-getaround-masabi-cities-data. Accessed 28 Dec 2018
31. Jackson, O.: The Supply Side of Discrimination: Evidence from the Labor Supply of Boston Taxi Drivers. Federal Reserve Bank of Boston, Boston (2018). 18–2
32. Brown, A.E.: Ridehail revolution: ridehail travel and equity in Los Angeles. UCLA (2018)
33. Alemi, F., Circella, G., Handy, S., Mokhtarian, P.: What influences travelers to use Uber? Exploring the factors affecting the adoption of on-demand ride services in California. Travel Behav. Soc. **13**, 88–104 (2018). https://doi.org/10.1016/j.tbs.2018.06.002
34. Circella, G., Alemi, F.: Chapter five - transport policy in the era of ridehailing and other disruptive transportation technologies. In: Shiftan, Y., Kamargianni, M. (eds.) Advances in Transport Policy and Planning, vol. 1, pp. 119–144. Academic Press (2018)
35. Rayle, L., Dai, D., Chan, N., Cervero, R., Shaheen, S.: Just a better taxi? A survey-based comparison of taxis, transit, and ridesourcing services in San Francisco. Transp. Policy **45**, 168–178 (2016). https://doi.org/10.1016/j.tranpol.2015.10.004

36. Chauhan, S.: Ride hailing to ride failing: how viable is the Uber business model in 2020? | Forbes India Blog. Forbes India (2020). http://www.forbesindia.com/blog/digital-navigator/ride-hailing-to-ride-failing-how-viable-is-the-uber-business-model-in-2020/. Accessed 03 Feb 2020
37. Sühr, T., Biega, A.J., Zehlike, M., Gummadi, K.P., Chakraborty, A.: Two-sided fairness for repeated matchings in two-sided markets: a case study of a ride-hailing platform. In: Proceedings of the 25th ACM SIGKDD International Conference on Knowledge Discovery & Data Mining, Anchorage, AK, USA, pp. 3082–3092 (2019). https://doi.org/10.1145/3292500.3330793
38. Zoepf, S.M., Chen, S., Adu, P., Pozo, G.: The economics of ride-hailing: driver revenue, expenses and taxes. CEEPR WP, vol. 5 (2018)
39. Efrati, A.: Alphabet's waymo seeks outside investors. The Information (2019). https://www.theinformation.com/articles/alphabets-waymo-seeks-outside-investors. Accessed 28 Jan 2020
40. Lane, C.: Why electric cars still don't live up to the hype. Washington Post (2019). https://www.washingtonpost.com/opinions/why-electric-cars-still-dont-live-up-to-the-hype/2019/12/30/242ce200-2b29-11ea-bcd4-24597950008f_story.html. Accessed 28 Jan 2020
41. Higgins, T.: Tesla faces fresh challenges after hitting 2019 delivery goal. Wall Str. J. (2020). https://www.wsj.com/articles/tesla-cuts-price-of-china-made-model-3-cars-11578057829
42. Taub, E.A.: 10 Electric Vehicles to Watch. The New York Times, New York (2020)
43. Walker, J.: Why transit advocates just went to war with Elon Musk. CityLab (2019). https://www.citylab.com/transportation/2017/12/what-elon-musk-doesnt-get-about-urban-transit/548843/. Accessed 29 Jan 2020
44. Walmsley, J.: Watch out, Uber. Berlin is the new amazon for transportation (with lower fares). Forbes (2019). https://www.forbes.com/sites/juliewalmsley/2019/06/26/watch-out-uber-berlin-is-the-new-amazon-for-transportation-with-lower-fares/. Accessed 29 Jan 2020
45. Walker, A.: Free transit isn't enough. Transportation needs to be a right. Curbed, 20 December 2019. https://www.curbed.com/2019/12/20/21031126/free-transit-universal-transportation-access. Accessed 29 Jan 2020
46. Pimental: This Lyft partnership program may change city transportation. Bisnow (2018). https://www.bisnow.com/los-angeles/news/neighborhood/lyfts-partnership-program-with-monrovia-may-change-how-city-transportation-90408. Accessed 29 Jan 2020
47. Ruggiero: Lyft, Uber discounts with public transit extended in Dublin. East Bay Times, 08 July 2017. https://www.eastbaytimes.com/2017/07/08/lyft-uber-discounts-with-public-transportation-extended-in-dublin/. Accessed 29 Jan 2020
48. Cervero, R., Duncan, M.: Benefits of proximity to rail on housing markets: experiences in Santa Clara County. J. Public Transp. **5**(1) (2002). https://doi.org/10.5038/2375-0901.5.1.1
49. Cervero, R., Rood, T., Appleyard, B.: Tracking accessibility: employment and housing opportunities in the San Francisco Bay Area. Environ. Plan. A **31**(7), 1259–1278 (1999)
50. Fagnant, D.J., Kockelman, K.M.: The travel and environmental implications of shared autonomous vehicles, using agent-based model scenarios. Transp. Res. Part C Emerg. Technol. **40**, 1–13 (2014)
51. Guerra, E.: When autonomous cars take to the road. Planning **81**(5), 36–38 (2015)
52. Fulton, L., Mason, J., Meroux, D.: Three Revolutions in Urban Transportation. Institute for Transportation & Development Policy, Davis (2017)
53. Mekuria, M.C., Appleyard, B., Nixon, H.: Improving livability using green and active modes: a traffic stress level analysis of transit, bicycle, and pedestrian access and mobility (2017)
54. Riggs, W.W., Boswell, M.R.: No business as usual in an autonomous vehicle future (2016)

55. Riggs, W.: Technology, civic engagement and street science: hacking the future of participatory street design in the era of self-driving cars. In: Proceedings of the 19th Annual International Conference on Digital Government Research: Governance in the Data Age, Delft, Netherlands, pp. 4:1–4:6 (2018). https://doi.org/10.1145/3209281.3209383
56. Riggs, W., Ruhl, M., Rodier, C., Baumgardner, W.: Designing streets for autonomous vehicles. In: Road Vehicle Automation, vol. 6, pp. 111–122 (2019)
57. Schlossberg, M., Riggs, W.W., Millard-Ball, A., Shay, E.: Rethinking the street in an era of driverless cars. UrbanismNext (2018). https://urbanismnext.uoregon.edu/files/2018/01/Rethinking_Streets_AVs_012618-27hcyr6.pdf
58. Appleyard, B., Riggs, W.: "Doing the right things" before "doing things right": a conceptual transportation/land use framework for livability, sustainability, and equity in the era of autonomous vehicles. Presented at the Transportation Research Board 97th Annual Meeting Transportation Research Board, Washington, D.C. (2018)
59. Riggs, W., Boswell, M.R., Ross, R.: Streetplan: hacking streetmix for community-based outreach on the future of streets. Focus J. City Reg. Plan. Dep. Coll. Arch. Environ. Des. Calif. Polytech. State Univ. San Luis Obispo **13**, 59 (2016)
60. Pressman: Tesla's ride-sharing network could be a game changer (2018). https://evannex.com/blogs/news/teslas-ride-sharing-network-may-be-a-lucrative-opportunity. Accessed 29 Jan 2020
61. Korosec: Tesla plans to launch a robotaxi network in 2020. TechCrunch (2019). https://techcrunch.com/2019/04/22/tesla-plans-to-launch-a-robotaxi-network-in-2020/. Accessed 01 Feb 2020
62. FHWA: National household travel survey. National Household Travel Survey (2019). https://nhts.ornl.gov/. Accessed 01 Feb 2020
63. Price: Uber drivers keep just 50% of what you pay. Business Insider (2015). https://www.businessinsider.in/uber-drivers-keep-just-50-of-what-you-pay/articleshow/46145782.cms. Accessed 01 Feb 2020
64. Tesla: Introducing software version 10.0. Tesla, 26 September 2019. https://www.tesla.com/blog/introducing-software-version-10-0. Accessed 01 Feb 2020
65. Ammann, D.: The cruise origin story. Medium, 22 January 2020. https://medium.com/cruise/the-cruise-origin-story-b6e9ad4b47e5. Accessed 01 Feb 2020
66. Riggs, W.: Painting the fence: social norms as economic incentives to non-automotive travel behavior. Travel Behav. Soc. **7**, 26–33 (2017). https://doi.org/10.1016/j.tbs.2016.11.004
67. Chu, D.: Waymo one: a year of firsts. Waypoint - The official Waymo Blog (2019). https://blog.waymo.com/2019/12/waymo-one-year-of-firsts.html. Accessed 01 Feb 2020
68. Priddle: Cruise origin: 8 things to know about GM's driverless pod car. Motor Trend, 28 January 2020. https://www.motortrend.com/news/cruise-origin-gm-autonomous-car-pod-things-to-know/. Accessed 01 Feb 2020
69. Audi: The new Audi A8 – conditional automated at level 3. Audi MediaCenter (2017). https://www.audi-mediacenter.com:443/en/on-autopilot-into-the-future-the-audi-vision-of-autonomous-driving-9305/the-new-audi-a8-conditional-automated-at-level-3-9307. Accessed 01 Feb 2020
70. Mitchell, R., Wilson, S.: Beyond Uber, Volvo and Ford: other automakers' plans for self-driving vehicles. Los Angeles Times (2016). https://www.latimes.com/business/autos/la-fi-automakers-self-driving-20160819-snap-htmlstory.html. Accessed 01 Feb 2020
71. Walker, J.: The self-driving car timeline – predictions from the top 11 global automakers. Emerj (2019). https://emerj.com/ai-adoption-timelines/self-driving-car-timeline-themselves-top-11-automakers/. Accessed 01 Feb 2020

72. Chokshi: Mary Barra Says G.M. Is 'on Track' to roll out autonomous vehicles next year. The New York Times (2018). https://www.nytimes.com/2018/11/01/business/dealbook/barra-gm-autonomous-vehicles.html. Accessed 01 Feb 2020
73. Rauwald, C.: BMW, daimler self-driving venture may attract fiat Chrysler. Bloomberg (2020). https://www.bloomberg.com/news/articles/2020-01-14/bmw-daimler-s-self-driving-venture-may-attract-fiat-chrysler. Accessed 01 Feb 2020
74. Ammann, D.: The next steps to scale start in San Francisco. Medium, 24 July 2019. https://medium.com/cruise/the-next-steps-to-scale-start-in-san-francisco-713315f3a142. Accessed 01 Feb 2020
75. Murray, C.: Automakers are rethinking the timetable for fully autonomous cars. Design News, 17 May 2019. https://www.designnews.com/electronics-test/automakers-are-rethinking-timetable-fully-autonomous-cars/93993798360804. Accessed 01 Feb 2020
76. Reader, R.: Ford CEO confession: we 'overestimated' the arrival of self-driving cars. Fast Company, 10 April 2019. https://www.fastcompany.com/90332941/ford-ceo-confession-we-overestimated-the-arrival-of-self-driving-cars. Accessed 01 Feb 2020
77. Korosec, K.: Waymo CTO on the company's past, present and what comes next. TechCrunch (2019). http://social.techcrunch.com/2019/02/08/waymo-cto-on-the-companys-past-present-and-what-comes-next/. Accessed 01 Feb 2020

Enabling Transportation Networks with Automated Vehicles: From Individual Vehicle Motion Control to Networked Fleet Management

Lily Elefteriadou[1], Blaine Leonard[2], Lili Du[1], Wei Ma[3], Jun Liu[4], Kuilin Zhang[5], Jiaqi Ma[6(✉)], Ziqi Song[7], Xiaopeng Li[8], and Sevgi Erdogan[9]

[1] University of Florida, Gainesville, USA
elefter@ce.ufl.edu, lili.du@essie.ufl.edu
[2] Utah Department of Transportation, Taylorsville, USA
bleonard@utah.gov
[3] Carnegie Mellon University, Pittsburgh, USA
weima@cmu.edu
[4] University of Alabama, Tuscaloosa, USA
jliu@eng.ua.edu
[5] Michigan Technological University, Houghton, USA
klzhang@mtu.edu
[6] University of Cincinnati, Cincinnati, USA
jiaqi.ma@uc.edu
[7] Utah State University, Logan, USA
ziqi.song@usu.edu
[8] University of South Florida, Tampa, USA
xiaopengli@usf.edu
[9] University of Maryland, College Park, USA
serdogan@umd.edu

Abstract. The technologies and models linking individual vehicle control and network operations, despite their critical role in determining whether automated vehicle (AV) technologies can eventually evolve to massive real-world deployments, seem to be an under-represented topic at the Automated Vehicles Symposium (AVS) in the past years. This chapter documents lecture notes of the first AVS breakout session on network modeling, which discusses the latest developments in network AV operations, modeling and simulation from academia, government, and industry perspectives. Specifically, the consensus reached in an attempt to answer questions on how to manage AV fleets in a networked environment, and control mixed traffic to optimally utilize network capacity and challenges ahead, as well as directions for research, practice and policy making are summarized.

Keywords: Vehicle control · Network modeling · AV fleet management · Network capacity

G. Meyer and S. Beiker (Eds.): AVS 2019, LNMOB, pp. 49–62, 2020.
https://doi.org/10.1007/978-3-030-52840-9_5

1 Introduction

The technologies and models linking individual vehicle control and network operations are critical in determining whether automated vehicle (AV) technologies can eventually evolve to massive real-world deployments. The breakout session at the AVS 2019, titled "Enabling transportation networks with automated vehicles: from individual vehicle motion control to networked fleet management", was organized to serve as a forum for gathering experts from academia, government and industry to discuss the potential opportunities and challenges for integrating individual vehicle technologies into transportation network modeling. This session brought together major AV fleet companies representing multiple modes (e.g., transportation network companies, transit, and trucks) to exchange questions and visions with representatives from governments and academia to identify the common issues impeding them from network-wide deployment, such as individual motion control technology maturity, path and route planning challenges, network fleet management, and mixed traffic environment.

The session aimed to provide an understanding of existing challenges and explore future solutions to network-wide deployment, taking advantage of the great mass and capacities of transportation network modeling communities. The session assisted researchers, government agencies, and industry in characterizing the needs and challenges in shaping future cities and identifying pathways towards the materialization of advanced transportation technologies.

This chapter provides a high-level summary of lectures by the invited speakers at the breakout session on network modeling methods and data need. The chapter starts with presenting latest efforts on vehicle-level modeling and testing with CAV pilots in Sect. 2. Then, novel network modeling frameworks for connectivity and automation are covered in Sect. 3 followed by a discussion of emerging data technologies and modeling methodologies as the enabler for future CAV network modeling in Sect. 4. Finally, concluding remarks from the discussions and directions for future research are summarized in Sect. 5.

2 Vehicle-Level Modeling and Testing with CAV Pilots

2.1 Using Connected Transit Data to Assess Traffic Impacts: Experiences from the Utah Testbed

In November 2017, the Utah Department of Transportation (UDOT) deployed an operational connected vehicle corridor in the Salt Lake City area involving Utah Transit Authority (UTA) buses [1]. In this deployment, a select number of buses and traffic signals along the corridor were outfitted with dedicated short range communication (DSRC) radios which allowed the buses to request signal priority at intersections when they were running at least 5 min behind schedule. This deployment may be the first vehicle-to-infrastructure (V2I) deployment in the United States using DSRC technology as part of an operational transportation system. In addition, this deployment provided a unique opportunity to use actual field data to assess the performance of a transit signal priority (TSP) system, which has typically only been possible with models.

The site of this deployment is Redwood Road, a state-owned, north-south arterial just west of downtown Salt Lake City, Utah. The portion of Redwood Road selected for installation of connected vehicle technology is about 11 miles long with 30 signalized intersections, extending from 400 South to 8040 South. Evaluation of transit schedule performance in this study is focused on 6 miles of the corridor. DSRC radios were installed on 25 of the 30 Redwood Road intersections, and on four UTA buses, along with single-board Linux computers to operate the software systems.

Roadside software developed for this project pulls data from the traffic signal controller, and broadcasts, through the DSRC radio, signal phase and timing data (SPaT) and a digital representation of the intersection geometry (MAP). When the bus approaches an intersection, on-board systems broadcast a "basic safety message" (BSM) to report the location, speed and direction of the bus. The on-board software also queries the UTA mobile data computer to determine the on-time status and the occupancy of the bus and, using the MAP data, determines if the bus is in a through lane. When the bus is more than 5 min behind schedule and has at least 20% occupancy, the system sends a "signal request message" (SRM) through the DSRC radios to the roadside, requesting priority (extra green time) at the intersection. Priority is granted by the signal controller unless other activities prevent that action. Prior to this deployment, buses along Redwood Road arrived at their stops on time about 86% of the time. The goal for this project was to increase that performance to 94% with minimal impact to surrounding traffic.

Data for the evaluation of bus performance came from three sources: 1) the messages sent through the connected vehicle system (BSM, SPaT, MAP, and SRM), 2) high-resolution signal controller data, including information about whether priority was granted when requested, and 3) bus reliability and occupancy data from the UTA mobile data computer. Only through the synthesis of this field data was the evaluation of bus performance possible.

Evaluation of the data indicated that buses which were equipped with the DSRC-based TSP system arrived at key time points along the route six percent more often during peak travel times than buses without the system. This improved performance met the goals of the project. Further, the analysis indicated that delays for traffic crossing the intersections where buses were requesting priority were infrequent. This is largely due to the fact that when buses request priority, modifications to the signal timing scheme at that intersection occur infrequently (about 10% of the time) because the bus is often able to pass through the intersection during the normal "green" phase. In that case, the priority request is then cancelled.

The TSP system using connected vehicle technology has been shown to provide tangible benefits without significant negative impacts on other vehicles. The system is being expanded to other corridors, and further performance studies are underway. Further, this study demonstrated that the ability to communicate between vehicles and the infrastructure provides data and tools which enable us to optimize network performance and capacity in ways that are not possible without this communication.

2.2 Capitalizing on Autonomy and Connectivity to Enhance Mobility: The I-STREET Testbed

The University of Florida (UF) and its Transportation Institute (UFTI - http://www.transportation.institute.ufl.edu/), along with the Florida Department of Transportation (FDOT) and the City of Gainesville (CoG) have partnered in the development of a real-world smart testbed on the highway network around the UF campus. The testbed, named I-STREET (Implementing Solutions from Transportation Research and Evaluation of Emerging Technologies, https://www.transportation.institute.ufl.edu/research-2/istreet-about-us/) has been deploying and evaluating numerous advanced technologies including connected and autonomous vehicles, smart devices, and sensors to improve mobility and safety. These technologies and their application are designed to work within the existing network and accommodate the presence of conventional vehicles. Industry partnerships are being sought to facilitate the development and operation of the testbed.

The following are examples of projects currently underway:

- **AV shuttle.** It operates in the downtown Gainesville area, under a pilot project funded by FDOT. The shuttle is operated by the Regional Transit System (RTS) of Gainesville, and it is available for free to the general public. An on-going project is evaluating the acceptance and use of autonomy, and the behavior of traffic around the AV shuttle.
- **Gainesville SPaT Trapezium.** This project is deployed with connected vehicle (CV) technologies and applications along four roads forming a trapezium surrounding the University of Florida main campus. The routes are SR 121 (SW 34th St), SR 26 (W University Ave), US 441 (SW 13th St), and SR 24 (SW Archer Rd). The goal of the project is to improve travel time reliability, safety, throughput, and traveler information. This project will also deploy and test pedestrian and bicyclist safety CV and smartphone-based applications. The project covers 27 traffic signals equipped with 27 Roadside Units. The project became operational in September 2019.
- **Optimal AV/CV trajectories at Signalized Intersections.** This initiative has been funded by NSF and FDOT, and it has developed, tested, and deployed an intelligent real-time intersection traffic control system that can simultaneously optimize signal control and AV trajectories, considering the presence of autonomous, connected, and conventional vehicles in the traffic stream.
- **Bicycle rack sensors on buses.** In this project, UF researchers have developed and deployed sensors on bike racks located in front of transit buses, which detect whether the racks are occupied. An app was also developed to provide real-time bike rack availability information to travelers interested in traveling by bus with their bicycles.
- **I-75 FRAME (Florida's Regional Advanced Mobility Elements.** This project deploys CV technologies to better manage, operate, and maintain the multi-modal transportation system and create an Integrated Corridor Management solution along I-75 and state highway systems in the Cities of Gainesville and Ocala. Applications include traffic signal control, transit signal priority, freight signal priority, and real-time traveler information.

Several additional projects are currently underway or being planned (for example, a data analytics platform, transit bus safety through advanced driver assistance systems, and school zone safety application). Industry partners have been engaged to participate in I-STREET, and the stakeholders are continuously evaluating the results of the tests as well as findings from other testbeds to plan future research.

3 Network Modeling Framework for Connectivity and Automation

3.1 Coordinated In-Vehicle Routing Built upon Online Learning and Distributed Optimization for Connected and Autonomous Vehicles

Among a number of approaches for in-vehicle routing, there are two groups that are rather accessible to both academia and practitioners. One is independent routing, which simply disseminates the information of instantaneous traffic conditions of an interested network to equipped vehicles, and expects each driver to independently make his/her own route choice. The second is systemic routing, which collects all drivers' origin-destination information for a centralized decision unit to systemically make an overall route decision for all involved vehicles. It is well known that independent routing leads to selfish routing and results in oscillated traffic conditions in the network, while systemic routing is for the best interest of the whole network, but not necessarily individual vehicles. Moreover, the computational load in the second approach is too high to be feasible for a real application. To address the dilemma between the above two groups of approaches, this study proposes the third in-vehicle routing mechanism: a novel coordinated online in-vehicle routing mechanism (CRM), assuming smart vehicles equipped with wireless communication and local computation facilities.

Briefly, the CRM is run by the following procedure. Step 1: Using onboard-connected vehicle communication devices, each vehicle in the coordination group sends its tentative best routing decision (route choice or routing preference) to the traffic center based on the current traffic condition. Step 2: The traffic center updates traffic information taking these routing decisions into account, and sends back out to the CVs. Step 3: CVs use this new information to update their best routing decisions. Step 2 and Step 3 repeat and continue until no CV is willing to change its routing decision upon receiving new traffic information (i.e., equilibrium routing decisions among vehicles). A single user can use this route guidance tool multiple times throughout a single trip as needed.

The iterative process of CVs updating their route decisions and communicating to the traffic center is handled internally using the CVs computation and communication capabilities (requiring only some general initial routing preferences as input from the human driver). We expect that the coordinated route decision—which considers the route decisions of other CVs in the coordination group—helps mitigate the traffic congestion resulting from individual CVs' overreaction[1] to real-time traffic information. Please note that the CRM works on the in-vehicle routings of a group of vehicles rather than only guiding an individual vehicle. Thus the CRM seeks to balance user optimality and system optimality.

The CRM models the routing decision process of a group of smart vehicles as either a pure strategy or a mixed strategy routing game, in which smart vehicles decide their own online shortest path or route choice priorities by a negotiation and coordination process with other smart vehicles. We label these two types of CRM as CRM-P and CRM-M respectively. As a mixed strategy game is employed, a discrete choice model is employed to count for drivers' behavior for the CRM-M. Our study [2, 5] shows the existence of an equilibrium coordinated routing decision for the CRM-M. Furthermore, a simultaneously updating distributed algorithm is proposed to implement the CRM-M. And, the convergence of the distributed algorithm to the equilibrium routing decision is proved, assuming individual smart vehicles are selfish players seeking to minimize their own travel time. As a pure strategy based game is used, this study [3, 4] also proves the existence of the equilibrium routing decisions of the CRM-P. Accordingly, a sequentially updated distributed algorithm is developed and its convergence to an equilibrium solution of the CRM-P is also proved. Numerical experiments conducted based on Sioux Falls city network indicate that the two distributed algorithms converge quickly under different smart vehicle penetration levels, thus they both possess a great potential for online applications. Moreover, the proposed (either pure strategy based or mix strategy based) coordinated routing mechanisms outperforms traditional independent selfish-routing mechanism; it reduces travel time for both overall system and individual vehicles, which represents the core idea of Intelligent Transportation Systems.

Furthermore, our experimental results show that the CRM-P will lose the computation efficiency if more than 4000 travelers en route are involved in a CRM. Our study in-depth also noticed that travelers will bring in small benefits in the CRM-P/M for mitigating traffic congestion if their routes are not correlated to other travelers in the coordination group. Accordingly, to scale up the applicability of the CRMs in a big city, we propose to implement CRM for multiple coordination groups (CG) each with a limited number of selected travelers. We label this type of CRM as CG-CRM. It can be integrated into either the CRM-P or CRM-M. Machine learning approaches are developed to find the CGs online according to the travelers' origin-destination and their candidate path sets [6]. Our numerical experiments show that the CG-CRM still outperforms the independent routing mechanism. It can significantly benefit computation efficiency with minor system performance loss as compared to the CRMs built upon a single group involving all smart vehicles.

Last, having recognized the price of anarchy issue for an equilibrium solution of a game model, we developed an information perturbation (IP) strategy which is inserted into Step 2 of the CRM to influence individual users' routing decisions and lead an equilibrium routing resolution to approach the desired level of system optimality. This

[1] Each CV independently does the best response (such as chooses the shortest path) according to the real-time traffic information without routing coordination.

gives rise to two new CRM schemes: CRM-M-IP [7] and CRM-P-IP [3]. Our theoretical studies provide the theoretical guideline to strategically implement the IP strategy with the merit of balancing system gain (i.e., reducing total system travel time) and user loss (a near-shortest path with longer travel time). Both theoretical analysis and the numerical experiments show that a small information perturbation following the proposed instruction will lead to significant system gain under either the CRM-M-IP or CRM-P-IP. Information perturbation is a promising strategy to ensure desired system performance, while also counts users' compliance.

Overall, the CRMs consider a smart transportation system (such as AV transportation system) as a dynamic multi-agent system and creatively influence individual vehicles' route choices so that the desired collective system performance, considering traffic network-level mobility and environmental sustainability can be achieved.

3.2 Challenging Human Driver Taxis with Shared Autonomous Vehicles: A Case Study of Chicago

Before autonomous vehicles are ready for daily use by the public, the transportation community must plan for their ultimate deployment in a logical, incremental manner. Sharing autonomous vehicles, as an "extension" of the current car-sharing concept, appears to a promising travel mode in the future with self-driving cars. Car-sharing offers travelers a mode without the burden of owning a vehicle [8, 10]. In comparison with current car-sharing programs where human drivers are necessary, shared autonomous vehicles (SAVs) can bring competitive advantages to car-sharing services. SAVs do not require travelers to make a trip to access a vehicle or after returning a vehicle (e.g., Car2Go). SAVs can drive themselves to locations requested by travelers. The cost of using SAVs does not include the cost of driver labor which services provided by Uber, Lyft, or taxis must bear and pass on to consumers. Travelers who do not own a personal vehicle, do not have access to a private car (e.g., visiting a non-home city), or are unable to drive are expected to be the first SAV riders. Indeed, it has been proposed that current taxi users are expected to be frequent SAV users [11, 12]. As such, current taxi services with human drivers may not be able to survive when a fleet of SAVs are on the road. This study envisions a city with SAVs that challenge human driver taxi services. The study is based on an actual data set of some 1,703,000 taxi trips. It applies agent-based models to simulate various SAV deployment scenarios within the network to test the influences of prices and operational schemes on mode choice between SAVs and human driver taxis. The results are intended to provide insights to both private and public sector entities seeking to implement SAVs in their cities to challenge or eventually replace existing taxis.

This study used the taxi trip data, downloaded from the data portal of the City of Chicago [13] to simulate the operations of SAVs competing with human driver taxis. This study looks into taxi trips recorded in 2014, before the dramatic decline in records due to the influences of ride-hailing services such as Uber and Lyft [14]. The data do not include trips from ride-hailing services. The data used in this study were error-checked. Records with incomplete or incorrect information were removed. Given that travel demand is periodic in nature, this study uses two days of taxi trips on May 14

(Wednesday) and 17 (Saturday), to represent the weekday and weekend travel demand. In total, there were 75,085 and 97,704 valid taxi trips from the two selected dates.

Figure 1 demonstrates the agent-based modeling process with mode choice to generate a fleet of SAVs serving travelers who often call taxis. At the time of trip departure, a traveler makes a riding request to SAVs or taxis. The search for vehicles informs the traveler whether there are available vehicles nearby. During the search, the traveler needs to make a choice between calling a taxi or SAV. The details of mode choice are provided below. If the traveler chooses to call an SAV, then the available closest SAV will go to the requested pickup location. If there is no available SAV within the range of a tolerable waiting time, a new SAV is generated at the pickup location and added into the fleet. Then the vehicle picks up the traveler, drives to the requested destination, and then drops off the traveler. At this point, this vehicle becomes available for next ride.

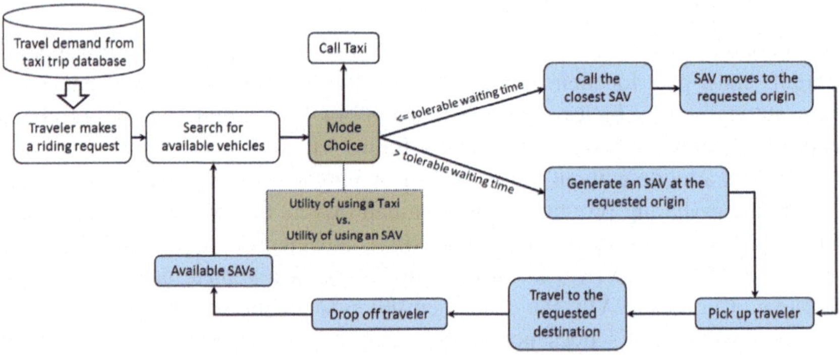

Fig. 1. Agent-based modeling process to generate a fleet of SAVs [8]

The mode choice is simulated by comparing travel utilities of using two modes: SAV and Taxi. A logit model is used to calculate the probability of choosing one mode over the other, according to the travel utilities. The probability of choosing SAV over taxi can be obtained:

$$P_{SAV} = \frac{\exp(U_{SAV})}{\exp(U_{SAV}) + \exp(U_{Taxi})} \quad (1)$$

Where, U_{SAV} = utility of using an SAV; and U_{Taxi} = utility of using a Taxi. The probability of using an SAV is:

$$P_{Taxi} = 1 - P_{SAV} \quad (2)$$

The probability shows the likelihood of a particular mode being chosen in a stochastic rather than deterministic simulation process. Since a driver is not required in SAV services, all driver-associated costs (e.g., driver salary), which can an important component of the taxi service cost, will not be a burden to SAV investors and operators. Simulations in this study imagined a few possible costs of using SAV services by subtracting the driver labor costs from the taxi payments. Especially, the costs of SAV services are equal to the amount of taxi charges subtracting 50%, 75%, 100% labor costs respectively. This sturdy implemented a stochastic process to simulate to mode choice between SAVs and taxis, and output mode splits when different SAV costs are given. The results showed that SAVs are called to serve 77 to 86% of trips. As expected, lower prices are certainly able to attract more travelers to choose SAVs, if all other factors are held constant. When the cases in simulations are true in the future, it may be very difficult for human driver taxis to survive when SAVs touch the ground.

SAVs with technological advances can bring sustainability benefits to the environment [15–17]. However, the empty VMT may negate the benefits gained from SAVs' technological advances. Fortunately, based on the simulation results in this study, empty VMT is not sizable enough to completely cancel out the sustainability benefits from technological advances. SAVs are still expected to save 9 to 16% fuels and produce 2 to 10% less GHG.

3.3 An Integrated Network Fleet and Routing Optimization Model for On-demand Shared Mobility Systems Using Connected and Automated Vehicles

CAV technologies and shared mobility concepts are foundations to develop the next generation innovative transportation services such as mobility-as-a-service (MaaS). For autonomous driving companies, there are very high costs of sensing and computing hardware and software of connectivity and automation such as LiDAR, high-precision GPS, and High-Definition Maps. On the other hand, the majority part of the revenue for transportation network companies pays for the cost of drivers. Therefore, both types of companies (such as Waymo and Uber) have recognized that the future feasible business model should provide CAV riding services as MaaS.

There are two key research questions to implement the CAV riding service system as follows. (1) How to automatically determine the fleet size of the CAVs and which CAV to be automatically assigned to serve a particular travel request? (2) How to automatically determine the optimal route and pickup/delivery decisions of travelers for these shared-use CAVs in a real-world network with vehicle capacity constraints. Essentially, the problem of interest is to find the minimum fleet size and transportation costs such that satisfy all the passenger travel requests in a transportation network.

This problem is related to two types of research fields in the literature. The first one is the minimum network fleet problem for on-demand mobility. For example, Vazifeh et al. [18] use 150 million taxi trip data to address the minimum fleet problem in on-demand mobility utilizing a concept of vehicle-shareability network. The second one is the classical vehicle routing problem (VRP), or specifically, the pickup and delivery problem with time window (PDPTW). For example, Mahmoudi and Zhou [19] provide a state-space-time network-based formation for the optimization of on-demand transportation systems and ride-sharing services. However, these models still target strategic and tactical level decisions, not operational decisions of autonomous vehicles that require real-time decisions on real-world roads such as automatically determining locations at curbside pickup/drop-off multiple travelers simultaneously.

The challenge for this operational decision-making problem is to determine a set of multidimensional integer decisions such as determining the optimal network fleet size, matching travel requests with vehicles, selecting vehicle routes, and pooling passengers simultaneously by considering classical routing and assignment constraints as well as time connectivity and vehicle dynamic pickup/drop-off and capacity constraints. The problem is difficult since all the decisions are multidimensional integer decisions, and many constraints are nonlinear integer as well. To be able to solve this problem using off-the-shelf commercial solvers, there is a need to develop effective model reformulations to deal with the nonlinear constraints. Also, efficient algorithms using decomposition methods should be developed for solving large-scale problems. Another difficulty of the problem is the demand uncertainty since the travel requests are hard to be predicted in real-time due to the uncertainty in both temporal and spatial dimensions. This problem is still a challenge for existing transportation network companies. Data-driven methods such as machine learning may help provide a better prediction for real-world applications.

4 Emerging Data Technologies and Modeling Methodologies

Estimating the fundamental traffic state variables, namely, flow, density, and speed plays a fundamental role in traffic engineering. Conventional traffic state estimation methods rely on fixed traffic sensors such as loop detectors, cameras and microwave vehicle detectors. Due to the high cost of conventional traffic sensors, traffic state data are usually obtained in a low-frequency and sparse manner. In contrast, the last decades have witnessed the breakthrough of automated vehicles (AVs). It is projected that more and more AVs will operate on roads in the next several decades, and various sensors installed on the AVs, including, but are not limited to, LiDAR, radar, camera and stereovision, will be collecting massive data and perceiving the surrounding traffic states. In fact, a fleet of AVs can serve as floating (or probe) sensors, which can be utilized to infer traffic information while cruising around the roadway network, as presented in Fig. 2. In view of this, this study builds a two-step framework that leverages the sensing power of AVs to estimate high-resolution traffic state variables. The first step directly translates the information observed by AVs and the second step employs data-driven methods (e.g., matrix completion, regression) to estimate the information that is not observed by AVs. The proposed estimation methods are data-driven and can be interpreted by the traffic flow theory.

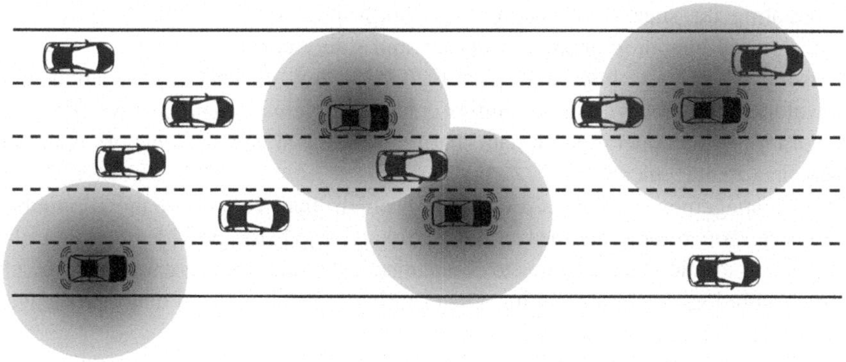

Fig. 2. An illustration of the traffic sensing framework with AVs.

The Next Generation Simulation (NGSIM) data is adopted to examine the accuracy and robustness of the proposed framework. We assume that a random set of vehicles are AVs and the AVs can perceive the surrounding traffic conditions. Given the fragmented information collected by AVs, we estimate the traffic states using the proposed framework. Experimental results are compelling, satisfactory and interpretable. Sensitivity analysis regarding AV penetration rate, sensor configuration, and perception accuracy are also studied.

In addition to traffic states, we are able to detect and track various objects in transportation networks by leveraging the rich data collected through AVs. The objects include, but are not limited to, moving vehicles by vehicle classifications, parked vehicles, pedestrians, cyclists, signage in public space. When all those objects in high spatio-temporal resolutions are being continuously tracked, those data can be translated to useful traffic information for public policies and decision making. This study for traffic sensing also implies the future possibility of interventions for effective and timely traffic management. It enables real-time traffic monitoring, potentially safer traffic operation, faster emergency response, and smarter infrastructure management. We hope this study could help policymakers and private sectors (e.g., Uber, Waymo) to understand the values of AVs in traffic operation and management, and further promote the collaboration between private sectors and public agencies.

5 Conclusions

This chapter presented the lecture notes from the AVS 2019 breakout session titled "Enabling transportation networks with automated vehicles: from individual vehicle motion control to networked fleet management". The session objective was to assist researchers, government agencies, and industry in characterizing the needs and challenges in shaping future cities and identify pathways towards the materialization of advanced transportation technologies.

The following summarizes some of the challenges and needs identified during the two panel discussions as well as audience questions:

- Evaluating the success of autonomous shuttle pilot studies is challenging due to e.g., limited operation areas, low speeds and safety concerns.
- Sensing is a very important part of CAVs but dependency to applications developed/owned by multiple entities creates problems in data sharing. Thus, collaboration and cooperation on data sharing is a critical need for network operations and fleet management.
- Fleet management is seen as a complex problem that require collaboration among industry, government and academia. For example, HERE provides real-time data, maps and solution technologies to companies and works with researchers to develop new technologies and algorithms. Data sharing is identified as a major challenge in fleet management as well.
- Another challenge was modelling the mixed traffic: the consensus was that it will not be seamless and easy as there exists many unknowns some of which are: information display, interactions between human driven and AVs, how to enable AV learning in mixed traffics,
- Behavioral questions exist, e.g., despite the expected benefits on road capacity and efficiency, depending on the business models, congestion and capacity issues may still remain if AVs encourage more driving. Thus, implications on transit and future of autonomous transit requires further consideration.
- How policy makers will decide, how AVs will be used, what business model will take momentum is critical.
- Multimodality will remain as a key component. AVs will not end multimodality but they need to address reducing inefficiencies.
- Another major challenge and research need is on computational issues as network level problems are very complex. Research will be needed for scaling up exiting models which will require advances in hardware, software and advanced computational algorithm designs.
- More research is needed, combining data and models available in large scale networks so that impacts of AVs can be investigated.

In addition to these challenges and questions, the breakout session also identified a list of topic areas that requires attention from the research community to understand and enable network-wide AV deployments and operations.

- Network fleet management and optimization for shared mobility systems.
- Network traffic management and control with automated vehicles.
- Innovative network modeling methods that consider unique AV fleet characteristics.
- New demand modeling (activity and travel behavior) with AV as an alternative mode.
- Integrated multi-resolution analysis, modeling, computation and simulation tools for network AV traffic management and evaluation.

References

1. Demonstrating Transit Schedule Benefits with a DSRC-Based Connected Vehicle System: Transportation Research Record, Washington, D.C. (July 2019). https://trid.trb.org/view/1635543
2. Lili, D., Gong, S.: Stochastic poisson game for an online decentralized and coordinated parking mechanism. Transp. Res. Part B: Methodol. **87**, 44–63 (2016)
3. Lili, D., Han, L., Chen, S.: Coordinated online in-vehicle routing balancing user optimality and system optimality through information perturbation. Transp. Res. Part B: Methodol. **79**, 121–133 (2015)
4. Lili, D., Chen, S., Han, L.: Coordinated online in-vehicle navigation guidance based on routing game theory. Transp. Res. Rec.: J. Transp. Res. Board **2497**, 106–116 (2015)
5. Lili, D., Han, L., Li, X.: Distributed coordinated in-vehicle online routing using mixed strategy congestion game. Transp. Res. Part B: Methodol. **67**, 1–17 (2014)
6. Peng, W., Du, L.: Clustering based online coordinated in-vehicle routing built upon understanding the competition potential among travelers on network route resources. In: 98th Annual Meeting of the Transportation Research Board, Washington DC, January 13–17 (2019)
7. Spana, S., Du, L., Yin, Y.: An online in-vehicle coordinated routing mechanism for connected vehicles with information perturbation under mixed-strategy congestion game. In: 99th Annual Meeting of the Transportation Research Board, Washington DC, January 12–16 (2020)
8. Liu, J., Jones, S., Adanu, E.K.: Challenging human driver taxis with shared autonomous vehicles: a case study of Chicago. Transp. Lett. 1–5 (2019). https://doi.org/10.1080/19427867.2019.1694202
9. Liu, J., Kockelman, K.M., Boesch, P.M., Ciari, F.: Tracking a system of shared autonomous vehicles across the Austin, Texas network using agent-based simulation. Transportation **44** (6), 1261–1278 (2017)
10. Litman, T.: Autonomous vehicle implementation predictions. Victoria Transport Policy Institute (2017). https://www.vtpi.org/avip.pdf. Accessed 29 Apr 2018
11. Fagnant, D.J., Kockelman, K.M.: The travel and environmental implications of shared autonomous vehicles, using agent-based model scenarios. Transp. Res. Part C: Emerg. Technol. **40**, 1–13 (2014)
12. Shaheen, S., Galczynski, M.: Autonomous carsharing/taxi pathways. UC Berkeley (2014). https://itspubs.ucdavis.edu/files/carsharing_taxi.pdf. Accessed 30 Apr 2018
13. Chicago Data Portal. https://data.cityofchicago.org/. Accessed 25 Mar 2018
14. Schneider, T.: Chicago's Public Taxi Data (2017). http://toddwschneider.com/posts/chicago-taxi-data/. Accessed 23 May 2018
15. Liu, J., Kockelman, K.M., Nichols, A.: Anticipating the emissions impacts of smoother driving by connected and autonomous vehicles, using the MOVES model. In: Transportation Research Board 96th Annual Meeting (2016). http://www.caee.utexas.edu/prof/kockelman/public_html/TRB17CAVEmissions.pdf. Accessed 30 Jan 2017
16. Liu, J., Khattak, A., Wang, X.: The role of alternative fuel vehicles: using behavioral and sensor data to model hierarchies in travel. Transp. Res. Part C: Emerg. Technol. **55**, 379–392 (2015)
17. Loeb, B., Kockelman, K.M., Liu, J.: Shared autonomous electric vehicle (SAEV) operations across the Austin, Texas network with charging infrastructure decisions. Transp. Res. Part C: Emerg. Technol. **89**, 222–233 (2018)

18. Vazifeh, M.M., Santi, P., Resta, G., Strogatz, S.H., Ratti, C.: Addressing the minimum fleet problem in on-demand urban mobility. Nature **557**(7706), 534–538 (2018)
19. Mahmoudi, M., Zhou, X.: Finding optimal solutions for vehicle routing problem with pickup and delivery services with time windows: a dynamic programming approach based on state–space–time network representations. Transp. Res. Part B: Methodol. **89**, 19–42 (2016)

Catching up with Low-Speed Autonomous Shuttles

Katherine Turnbull[1(✉)], Cynthia Jones[2], and Lily Elefteriadou[3]

[1] Texas A&M Transportation Institute,
3135 TAMU, College Station, Texas, USA
k-turnbull@tti.tamu.edu
[2] DriveOhio, 1610 W. Broad Street, Columbus, OH, USA
Cynthia.jones@drive.ohio.gov
[3] University of Florida Transportation Institute,
365 Weil Hall, Box 116580 Gainesville, FL, USA
elfter@ce.ufl.edu

Abstract. This chapter presents information on low-speed autonomous shuttles, which are being piloted, demonstrated, and deployed in downtown areas, university campuses, business parks, entertainment complexes, and other areas. The chapter focuses on the presentations and discussions at a breakout session at the 2019 Automated Vehicle Symposium (AVS). Updated information on some demonstrations and projects is included. The session and this chapter highlight the experience with pilots and demonstrations and present elements to consider in planning, procuring, operating, and evaluating low-speed autonomous shuttles to help inform future decision-making. Areas for additional research and ongoing information sharing are also summarized.

Keywords: Low-speed autonomous shuttles · Driverless shuttles · Automated shuttles · Autonomous shuttles

1 Introduction

Building on the experience with automated public transport pilots in Europe as part of the European Commission–funded CityMobil2 project, numerous pilots and demonstrations of low-speed autonomous shuttles are being conducted in the United States. These activities range from one-day showcases to multi-month or year-long demonstrations to ongoing deployment programs. Initial projects included the Milo Driverless shuttle in the Arlington, Texas, Entertainment District, the Hop On Driverless Shuttle in Las Vegas, Nevada, and the Mcity Driverless Shuttle Research Project on the University of Michigan's North Campus. The AVS 2019 breakout group highlighted additional pilots and projects, some of which are summarized in this chapter, along with suggestions for planning, procuring, operating, and evaluating projects.

G. Meyer and S. Beiker (Eds.): AVS 2019, LNMOB, pp. 63–70, 2020.
https://doi.org/10.1007/978-3-030-52840-9_6

2 Examples of Low-Speed Autonomous Shuttle Pilots, Demonstrations, and Deployments

2.1 Bishop Ranch Shared Autonomous Vehicle Pilot Program

The Contra Costa Transportation Authority (CCTA) Shared Autonomous Vehicle (SAV) Program includes three phases covering the time period from 2016 to 2020. The project represents a partnership of the CCTA, GoMentum Station, Bishop Ranch, and EasyMile.

Phase I in 2016 included procurement of the EasyMile vehicles and obtaining federal and state approval for operating the SAVs on public roads. Initial testing of the SAVs was conducted at GoMentum Station. Phase II (2017–2018) included the first testing of an autonomous shuttle on public roads in Bishop Ranch. Testing of the shuttles uses trained operating personnel and Bishop Ranch employees scheduled as predetermined testers and evaluators. Phase III focuses on deploying SAVs on nine routes serving Bay Area Rapid Transit and bus stations throughout the county.

2.2 Columbus Smart Circuit

The Smart Circuit in downtown Columbus operated from December 2018 through September 2019, serving approximately 16,000 riders and operating over 19,118 miles. Smart Circuit represents a partnership involving Smart Columbus, DriveOhio, May Mobility, and Ohio State University. Three May Mobility autonomous shuttles operated along the Scioto Mile, serving the Center of Science and Industry, the National Veterans Memorial Museum, Bicentennial Park, and the Smart Columbus Experience Center. Service was provided from 6:00 a.m. to 10:00 p.m., seven days a week. The vehicles accommodated four passengers. A safety operator was onboard the vehicles at all times.

Other projects are planned to build on the experience with Smart Circuit. The One Linden project, which is part of the Columbus Smart City project, will use self-driving shuttles to close transportation gaps to connect public transportation, affordable housing, healthy food, childcare, recreation, and education. The route will provide access to and link St. Stephen's, the Douglas Community Center, the Rosewind Resident Council, and the Linden Transit Center. Other shuttle projects are being planned in Youngstown and Toledo. The SMART2 project in Youngstown includes autonomous transit shuttles and complete street enhancements. The Toledo project includes autonomous shuttles and apps to help locate available parking spaces.

2.3 Minnesota Winter Weather Testing and Demonstrations of Autonomous Shuttles

The Minnesota Department of Transportation (MnDOT) conducted winter weather testing of an EasyMile autonomous shuttle and held demonstrations of the vehicle in the Minneapolis-St. Paul Metropolitan area. The five project goals focused on snow and ice conditions, operations, mobility, infrastructure, influence, and partnerships. Public engagement was an overarching project goal. The five goals were: 1) prepare the autonomous vehicle industry for snow and ice conditions; 2) identify challenges and strategies for safe operation of third-party autonomous vehicles on MnDOT's transportation system; 3) identify the infrastructure that is needed to ensure safe operation of autonomous vehicles; 4) increase Minnesota's visibility and influence on advancing autonomous and connected vehicles; and 5) enhance partnerships between government and the autonomous vehicle industry.

The project included three phases. Phase 1 focused on testing the autonomous shuttle under snow, ice, and cold weather at the MnROAD research facility located to the northwest of the Minneapolis-St. Paul metropolitan area. Phase 2 included operating the autonomous shuttle on the Nicollete Mall in downtown Minneapolis during Super Bowl week in January 2018. Additional tests and demonstrations were conducted during Phase 3, including hosting members of the Minnesota Chapter of the National Federation for the Blind and other groups. Phase 3 activities also focused on investigating possible public and private partnerships for operating autonomous shuttles.

The demonstration plan developed for Phase 1 outlined the testing scenarios, the schedule, and the responsibilities of MnDOT, EasyMile, and the consulting team. The demonstration scenarios included obstacles, other vehicles, pedestrians, and bicyclists. Other scenarios focused on transit stops, stopping and yielding, and intersections. The scenarios were tested under fair weather conditions and winter weather conditions, including snow, ice, and road salt.

The public riding the autonomous shuttle on the Nicollet Mall during Super Bowl week and at other events were asked to provide feedback via text messages. The overall feedback was very positive. A report on the project is available [1].

2.4 Frisco drive.ai Pilot

This demonstration was sponsored by the Frisco Texas Transportation Management Association (TMA), which includes public- and private-sector members the City of Frisco, Denton County Transit Authority, Hall Park, the Star—home to the Dallas Cowboys, a 91-acre campus—and Frisco Station. The Frisco TMA focuses on developing the next-generation transportation solutions, promoting collaboration, and accelerating deployment.

The project was the first automated vehicle deployment with passenger service on public roadways in Texas. The eight-month deployment covered the time period from July 30, 2018 to March 29, 2019. Service was limited to employees and residents in the TMA area, with over 2,500 trips and almost 5,000 riders during the 8-month period. Community Days provided the public with the opportunity to ride the shuttle.

The experience from the Frisco drive.ai pilot showed that deployment of autonomous shuttles can be done safely. It also supported the importance of coordination for a safe deployment. Examples of coordination include preparing for possible incidents with first responders and public-relations personnel, conducting tabletop exercises with first responders, and providing first responders with access to the vehicles to help them understand their operation. The project also highlighted the importance of providing public education and outreach.

2.5 Denver Regional Transit District 61AV Project

The Denver Regional Transit District (RTD) conducted the six-month 61AV project from January through August 2019 as a first- and last-mile solution using an autonomous vehicle. The project connected RTD's 61st and Pena rail station to the Panasonic building, an emerging apartment complex and the 61st and Peña park-and-ride facility. The demonstration goals were to safely introduce an autonomous vehicle on a public roadway in Denver and to assess the reliability and availability of an autonomous shuttle vehicle and its suitability for a transit application. Other goals were to provide first- and last-mile service to and from an RTD bus/rail station and to align the interests of all stakeholders and partners to advance the project.

In addition to RTD, other partners included EasyMile, the autonomous vehicle provider, and Transdev, the operator and onboard customer service ambassador (CSA). Panasonic and LC Fulenwider, the co-developers of Peña Station Next were other partners. The City and County of Denver were also partners. The pilot project successfully completed approval processes involving the National Highway Traffic Safety Administration (NHTSA), the State of Colorado AV Task Force, the City/County of Denver, and Denver International Airport.

An EasyMile EZ10 Generation 1 (Gen 1) vehicle was used in the project. The electric shuttle accommodated up to 15 passengers (6 seated and 9 standing). A CSA was onboard during operating hours. The scheduled service was from 10:00 a.m. to 6:00 p.m., Monday through Friday.

A report is available summarizing the demonstration experience [2]. The daily percentage of scheduled service actually operated in autonomous mode ranged from a low of 46% to a high of 99%. Factors affecting the autonomous service availability included snow, heavy rain, and steam rising from the street due to melting snow. Other factors included severe cold weather (the Gen1 vehicle has no heater) and temporary obstructions, such as construction equipment. Approximately 600 people rode the shuttle during the demonstration.

2.6 Texas A&M University Campus NAVYA Demonstrations and Downtown Bryan Driverless Shuttle

Two demonstrations using NAVYA shuttles were conducted on the Texas A&M University campus in College Station, Texas, through the coordinated efforts of the Texas A&M Transportation Institute and Texas A&M University Transportation Services. A one-day demonstration of a NAVYA shuttle was held at Texas A&M on November 6, 2018. The autonomous shuttle operated with a safety officer onboard on the plaza in front of Kyle Field, the university's football stadium. Approximately 100 people rode the shuttle during four hours of operation. The 87 passengers completing surveys reported very positive feedback to the shuttle, including strongly supporting it as a useful and safe mobility option on campus and indicating they would ride it in the future.

A second demonstration of a NAVYA shuttle at Texas A&M was conducted from on September 9 through November 15, 2019. This demonstration was branded as the Smartshuttle, with the NAVYA vehicle operating on a one-mile route on public roadways. The shuttle was also available for viewing in static displays during football games and other on-campus events. Pre-planning activities included obtaining NHTSA approval, first responder and student safety driver training, coordinating with campus and community representatives, and developing a marketing and outreach program.

Approximately 600 miles of service was safely operated, with 90% operating in autonomous mode. Experience was gained with the vehicle software and with operating in hot and humid conditions. Rider responses to the shuttle were positive with support for more demonstrations and routes.

Faculty in the Texas A&M University Department of Mechanical Engineering and the university's Unmanned Systems Laboratory have developed and piloted self-driving platforms in a variety of vehicles, including shuttles, trucks, and sedans. These self-driving vehicles have been pilot tested at The Texas A&M University System RELLIS Campus, at a hotel and golf club complex, on the Texas A&M campus, and in downtown Bryan, Texas. The downtown Bryan Self-Driving Trolley pilot operated in October and November 2018. Two self-driving trolleys, with a safety officer onboard, operated on a loop serving a parking garage, the Bryan Library, and commercial establishments in downtown. Surveys conducted of riders on the different pilots have provided positive responses to the driverless shuttles.

2.7 University District AV Project Houston, Texas

The University District AV project currently includes two phases. The first phase is a one-mile closed loop route on the Texas Southern University (TSU) campus. The second phase will connect the University of Houston (UH) with the light rail transit (LRT) Purple Line and LRT to a transit center. The project was scheduled to operate from June 2019 to May 2020.

The Metropolitan Transit Authority of Harris County (Houston METRO) is the project lead. Project members include the Houston-Galveston Area Council (H-GAC), Port Houston, the City of Houston, TSU, UH, the Texas Department of Transportation, and the Texas Innovation Alliance. First Transit, Inc., is the shuttle operator and EasyMile is the shuttle vehicle provider. An operator is onboard to ensure safe operations. The electric EasyMile 10 Gen 2 vehicle has a capacity of 12 passengers (6 seated and 6 standing) and provides access for passengers with reduced mobility.

Operation of the autonomous shuttle began on the TSU campus on June 19, 2019, with ridership increasing over the first month of operation. There has been strong public interest and engagement. There have been no incidents with the service, and the vehicle's electric charging capabilities have been tracking as expected, despite the high temperatures and humidity of summer weather in Houston.

3 Planning, Implementing, Operating, and Evaluating Low-Speed Autonomous Shuttles

3.1 Planning for Low-Speed Autonomous Shuttles

Public transportation use cases for low-speed autonomous vehicles include fixed routes, circulators, shuttles, first-mile and last-mile connections to bus and rail services, and paratransit services. The use cases may be operated in a variety of settings, including university campuses and educational facilities; health care and hospital complexes; employment centers; and entertainment, recreation, and retail venues. Airports, parking facilities, retirement communities, and military bases are other possible settings.

Possible operational domains focus on the type of right-of-way and the interaction with other road users. Exclusive off-street guideways and off-street multi-use paths provide two examples of rights-of-way used with autonomous shuttles. Other alternatives include on-street pathways with a dedicated shuttle lane; on-street pathways with the shuttle operating in a dedicated bus lane; and on-street mixed-traffic lanes. Examples of elements to consider in project planning include feasibility, risk assessment and mitigation, operations, safety, and legal and regulatory factors. Public and stakeholder involvement is also an important element throughout the planning and deployment process. Obtaining necessary waivers from NHTSA is also an important step. The Mcity Driverless Shuttle: A Case Study [3] provides a good overview of all the elements involved in planning, implementing, operating, and evaluating a project.

3.2 Vehicle and Service Procurement

Three possible procurement models are an agency purchasing and operating autonomous shuttles, an agency contracting with a vendor to operate and procure the vehicles, and a service operator with a vehicle partner. Factors to consider in assessing possible procurement methods include the type of project, the schedule, and vehicle availability. The status of vehicle technologies and desired vehicle functions may also be considered.

Risk provisions to consider in the procurement process include insurance, cyber security, and indemnification. Third-party coordination may include involving agency staff, the vehicle manufacturer and operator, risk departments, and stakeholder coordination.

3.3 Implementation and Operation

Elements to consider in implementing and operating low-speed shuttles include involving stakeholders, operating procedures, emergency procedures, and safety meetings. Operating procedures address the initial delivery of the shuttle, daily site setup and check, passenger ride procedures, and evaluation procedures. Other elements include the detailed schedule of events, VIP activities, and apparel requirements for personnel.

Stakeholders include both critical personnel and interested individuals. Defining the roles and responsibilities of all groups and individuals throughout planning, implementation, operating, and evaluating a project is important. In addition to transit personnel, individuals from risk management, emergency medical services (EMS), police and fire, and traffic operations are typically involved. Developing procedures in case of an accident, emergency, or other related event is important. Procedures for evacuating the shuttle, notifying emergency personnel, and taking other actions should be addressed. The operational plan developed for the initial NAVYA demonstration on the Texas A&M University Campus highlights these and other elements [4].

As with any new technology that must be embraced by the public to be successful, promoting shuttle benefits during pilot demonstrations through local media is important. Messaging implementation in a positive way—as well as preparing personnel and agencies for how to approach public discourse, especially with the media, is key to garnering public support for the future, when more ubiquitous deployments of advanced transportation technologies, including autonomous shuttles, occurs.

3.4 Evaluations

Evaluation programs and evaluation measures should focus on the goals of the autonomous shuttle project. Typical elements of an evaluation program begin with defining the mission and purpose of the service, the goals and objectives, and the performance measures. The next steps include data collection and analysis, performance reporting, and the final project evaluation.

Considering the evaluation program in advance of a demonstration enables the collection of useful data that map to the specific project objectives. Traditional transit metrics are important but need to be linked to shuttle-specific goals or targets. Examples of traditional transit metrics include service supplied and consumed, cost efficiency and effectiveness, and service effectiveness and efficiency. Other traditional measures are service quality, safety, and service availability. These measures may be expanded to address more specific metrics related to autonomous shuttle operations.

New measures for autonomous shuttle pilots may focus on communities served, collaboration among public and private sectors, data sharing, and operational features. Other possible new measures focus on customer and user perspectives, public acceptance, and rider satisfaction. The pilots may represent the public's first exposure to automated driving technologies and may influence public acceptance. Focus groups, surveys, and interviews are being used with some pilots to obtain feedback from users and the public.

4 Further Research

A number of areas for further research projects, pilots, and evaluation have been highlighted through projects, studies, and discussions at the AVS breakout sessions. Examples of topics for additional research include on-road and on-vehicle signing, common evaluation methodologies, core questions for user and public surveys, and workforce impacts. Other topics are sensor robustness and performance, remote supervision and monitoring, and accessibility for all users. Continuing to share experiences with pilots, demonstrations, and deployments will also be important.

References

1. WSB and AECOM: MnDOT Autonomous Bus Pilot Project Testing and Demonstration Summary, Minnesota Department of Transportation, St. Paul, Minnesota, June 2018
2. Memorandum to the RTA Board of Directors from Bruce Abel, 61AV-Demonstration Project, 6 August 2019
3. University of Michigan, Mcity Driverless Shuttle: A Case Study
4. Texas A&M Transportation Institute and Texas A&M University Transportation Services, NAVYA Technology Demonstration at Texas A&M University Operational Plan, 4 November 2018

Part III Users and Human Factors

Not So Autonomous Vehicles: A Path to Consumers' Changing World

Johanna Zmud[1(✉)], Ipek N. Sener[2], Barbara Lenz[3], and Victoriya Kolarova[3]

[1] Texas A&M Transportation Institute (TTI), 1747 Pennsylvania Avenue, Suite 400, 20006 Washington, DC, USA
j-zmud@tti.tamu.edu
[2] Texas A&M Transportation Institute (TTI), E 505 Hutland Dr. Suite 455, 78752 Austin, TX, USA
i-sener@tti.tamu.edu
[3] German Aerospace Center (DLR), Rudower Chaussee 7, 12489 Berlin, Germany
{Barbara.Lenz,Viktoriya.Kolarova}@dlr.de

Abstract. The main premise of this chapter is that researchers, policy makers, and industry can learn a lot about the future adoption (and impact) of fully automated, self-driving vehicles by examining consumer acceptance, understanding, and trust of lower levels of automation (SAE Level 1 and Level 2). These are the levels of vehicle technology that exist on vehicles operating on public roads today. The willingness (or not) of consumers to purchase and use the lower levels of automation will serve as a mirror on their potential interests in higher levels. This chapter provides a synthesis of issues and recommendations that resulted from a breakout session at the 2019 Automated Vehicle Symposium (AVS) on this topic.

Keywords: Levels of driving automation · Acceptance and adoption of automated vehicles · Trust in transportation technologies · Self-driving vehicles

1 Introduction

The focus of most automated vehicle (AV) research studies has typically been on the highest possible levels of automation (i.e., self-driving vehicles). This is because fully self-driving vehicles have the potential to radically alter peoples' lives, transform our transportation system, and impact the spatial structure of cities and regions. However, the path to consumers' acceptance, understanding, and trust of fully self-driving vehicles is likely influenced by their familiarity with the automated driving features on vehicles now. Such vehicles have been entering the market at increasing rates throughout the past few years. These lower-level automated driving systems take on some aspects of safety-critical control functions, such as lane keeping assists and collision avoidance systems, to avoid a crash or collision. Some functions go beyond safety issues and are also related to the drivers' comfort, such as vehicle distance control in stop-and-go traffic situations.

G. Meyer and S. Beiker (Eds.): AVS 2019, LNMOB, pp. 73–77, 2020.
https://doi.org/10.1007/978-3-030-52840-9_7

This chapter provides a synthesis of issues and recommendations that resulted from a breakout session at the 2019 Automated Vehicle Symposium (AVS) on this topic. The session consisted of two panels—each including a moderator and four presenters—followed by an interactive discussion between the panel and the audience. The focus of the first panel was on 'Policy and Consumer Education' and the second panel focused on 'Public Acceptance and Trust'.

2 SAE Levels of Driving Automation

SAE issued a taxonomy and definitions (J3016) in 2016 that defined six levels of driving automation, from no automation to full automation, in order to speed the delivery of an initial regulatory framework and to facilitate discussions among stakeholders in the AV technology community. SAE's levels of autonomy have become a de facto global standard. In 2018, SAE updated the way in which its taxonomy was visually displayed to offer more "consumer-friendly" terms and definitions for the levels of driving automation in an attempt to help eliminate confusion among consumers, media, and industry.

The latest definitions of SAE levels are based on the rhetorical question: Are you the driver or are you a passenger? At the lower levels of automation (denoted as SAE Levels 0, 1, and 2), "you" are the driver. Level 0 describes a vehicle with no automated assistance. Level 1 describes many of today's new cars, in which the human driver is responsible for the safety and operation at all times, but the car can take over at least one vital function: steering or speed control. Cruise control is one of the best examples of existing technology at this level. Today's more advanced vehicles are at Level 2. The driver is still responsible for the safe operation of the vehicle, but the vehicle can take over steering, braking, and acceleration under certain conditions. The driver is expected to do everything else and continuously monitor road conditions. Tesla's Autopilot and similar systems from Mercedes-Benz, BMW, and Volvo are examples of partial automation. Because consumers actually own vehicles at these lower levels of automation, researchers have been able to examine their acceptance, trust, and understanding of them.

Starting with SAE Level 3, "you" are a passenger unless the automated driving feature requests otherwise. Level 5 is the highest level of automation, where the automated driving feature can drive the vehicle under all conditions.

3 Issues

Today, there is definite interest in automated driving features among consumers. A 2018 Consumer Reports survey found that 57% of drivers reported that advanced driver-assist features in their vehicle had kept them from getting into a crash (Fisher 2019). According to a 2019 American Automobile Association (AAA) survey, 55% of Americans want AV technology in the next vehicle they buy or lease (Brannon 2019). But there is also a general confusion among drivers around what is an automated driving feature and what is self-driving technology. According to the same AAA survey, 40% expect Level 2 driver support systems with names like Autopilot, Pro-PILOT, or Pilot Assist to have the ability to drive the car by itself. Similarly, a 2019 Insurance Institute for Highway Safety (IIHS) study found misconception of several common Level 2 automated features on the market today (IIHS 2019). All Level 2 automated features require the driver to continuously stay engaged in the driving task, but that is not how drivers perceived such systems in practice. Similarly, a 2018 MIT AgeLab survey found that 23% of respondents thought self-driving vehicles were available for purchase today (Reimer 2019).

As addressed in the session, AV technologies hold the promise of improved safety, and many other benefits, but only if they are understood accurately by the consumers. That is why researchers, practitioners, and the media need to be attentive to how the AV terminologies are being used now and work collaboratively towards improvement and standardization.

The terminology used to market the systems may create an inaccurate impression of vehicle capabilities among consumers. Such misconceptions can put drivers and other road users at risk. There is no current standard for nomenclature between different makes, models or trim levels (e.g., there are as many as 40 terms for automatic emergency braking).

Vehicles available for purchase are becoming more complex and feature-rich, and at the same time, vehicle owners receive little to no training on how to use them. Many of these automated features are far from intuitive. Drivers may not have the understanding and skills necessary to successfully leverage the technologies, and technologists often assume ideal performance of both humans and automated driving systems (Reimer 2019). Research has shown that drivers often misuse the automated features. In National Safety Council (NSC) surveys, more than a third (37%) of drivers said they prefer to turn off safety features entirely because they are confusing, irritating, or give false positives; 40% said they've been startled or surprised by their vehicle (Nantel 2019). The results of a 2019 Transport Canada survey indicated confusion among a significant number of respondents in regard to the difference between features that provide a warning signal to the driver, versus those that assist with the driving task (e.g. forward collision warning versus automatic emergency braking) (Phillips 2019). According to the 2018 Consumer Reports survey, reasons given for turning off such automated features as lane keeping assist are: "the system is annoying", "it jerks the steering wheel which I find scary rather than reassuring", and "I sometimes fight with the system to keep the car where I want it." (Fisher 2019).

Owners of vehicles equipped with Level 1 and Level 2 technologies want to be better educated as to the technologies in their vehicles—61% of drivers polled by the National Safety Council say they wish they got more training about the proper use of safety features in their vehicles (Nantel 2019). Few people take the time to read an Owners' Manual. Transport Canada reports that only 24% of drivers of vehicles equipped with select Advanced Driver Assistance Systems (ADAS) technologies reported using the owner's manual to learn about these features (Phillips 2019). A general lack of knowledge regarding the lower levels of automated driving systems can lead to a continued lack of trust from the public. If the public lacks trust and confidence in the automated technology or finds them confusing and intimidating, they may be less likely to use them. In the 2019 Transport Canada survey, only 33% of respondents stated they would be comfortable riding in a fully automated vehicle (Phillips 2019).

4 Recommendations

Safety, that is getting to zero fatalities, is the main reason for automating vehicles. The use of vehicles at SAE Levels 1 and 2 is expected to reduce the rate of injurious and fatal motor-vehicle crashes; however, this result can only be realized if drivers can safely and effectively use these automated systems. Research indicates that many drivers have little knowledge about how to use them and even less experience actually using them. The presence of the automated technology could exacerbate negative incidents that may generally result from driving vehicles with unfamiliar technologies.

Therefore, it is necessary to educate consumers as to the limitations in automated features' capabilities and in correct use of existing automated features. It's important for drivers to understand: In any vehicle available for sale to the public today, the driver must remain in control and engaged in driving at all times. The stepwise numbers to Level 5 give a false sense that driver-assistance technology can do more than its functionality. Using "self-driving" or partially automated or similar terms to describe a vehicle that requires attention from the driver is likely to lead to even greater confusion and safety concerns than we have today (Nantel 2019).

Trust in automation erodes when vehicles do unexpected things. The marketing terminology used by the auto manufacturers is a big part of the problem. To end confusion, we need consistent, simple terms for driver-assistance systems. The updated guidance from SAE is a starting point. It attempts to communicate to consumers the principles of "I'm driving; I'm riding."

Who should educate consumers? It will take industry, engineers, auto dealers, non-government organizations, driving schools, and policymakers working together to ensure consumers learn what a vehicle is able to do and what it is not. More importantly, the drivers need to be aware of what he/she is responsible for and under what driving conditions. More research is needed to examine the relationship between acceptance and trust. Can acceptance occur without trust? Does acceptance mean trust? How do you manage expectations of vehicle owners as to the technology systems on their vehicles? Do systems do what people expect?

The discussions of this session, and thus this chapter, highlights the need for more attention to be focused on measuring consumer acceptance and on building understanding and trust of these lower levels. It also underlines the important role of collaborative effort in standardizing and maintaining an accurate understanding of the AV terminologies among the public.

References

Fisher, J.: Consumer experience with today's automated vehicles. In: Presented at the 2019 automated vehicle symposium 2019, Orlando, U.S. (2019)

Brannon, G.: Consumers uncertain about automated technology. In: Presented at the 2019 automated vehicle symposium 2019, Orlando, U.S. (2019)

Insurance Institute for Highway Safety (IIHS): New studies highlight driver confusion about automated systems. https://www.iihs.org/news/detail/new-studies-highlight-driver-confusion-about-automated-systems. Accessed 22 May 2020

Reimer, B.: Automation fueled vehicle features and implications for trust and acceptance. In: Presented at the 2019 automated vehicle symposium 2019, Orlando, U.S. (2019)

Nantel, K.: Policy and consumer education. In: Presented at the 2019 automated vehicle symposium 2019, Orlando, U.S. (2019)

Phillips, A.: Supporting awareness and confidence in automated vehicles. In: presented at the 2019 automated vehicle symposium 2019, Orlando, U.S. (2019)

Societal Expectations from Automated Road Mobility: Results of a Survey in Germany

Yasmin Halil(✉) and Gereon Meyer

VDI/VDE Innovation + Technik GmbH, Steinplatz 1, 10623 Berlin, Germany
{yasmin.halil, gereon.meyer}@vdivde-it.de

Abstract. This chapter summarizes the results of a survey among 1,000 interviewed persons that was conducted in Germany. The aim was to obtain a current picture of the population's expectations from self-driving (automated) vehicles. Twelve questions were asked to get an overview of the participants' opinions on the potential of self-driving vehicles regarding climate impact, time and monetary savings, the willingness to use automated vehicles, as well as their safety and risks. The results are compared to the current state of expert knowledge. The main finding is that the German population is skeptical towards self-driving vehicles. In comparison to a scientific perspective, the population is underestimating the potential and possible benefits of automated vehicles, while the risks are assessed similarly. In the conclusions, the findings of the studyare assessed in view of the COVID-19 pandemic.

Keywords: Automated vehicles · Road automation · Societal expectations · Future mobility · Public transportation systems · Impacts of automated driving

1 Introduction and Methodology

Efficient, comfortable, safe and green – there are high expectations for Automated Vehicles (AVs) in scientific and professional expert circles. However, the population's view on AVs is rather unclear. Therefore, a study on the topic "Automated driving in a Smart City" was conducted to obtain a contemporary understanding of how the population is feeling about automated vehicles. Within the study, a representative survey of 1,004 German residents was prepared and evaluated. The survey itself was run by *forsa Politik und Sozialforschung GmbH,* an established polling company in Germany. They performed computer-assisted telephone interviews (CATI) asking people to evaluate AVs with regard to their potential to increase the quality of life in cities, in terms of time and financial savings, and finally, regarding the reduction of traffic jams. Furthermore, the interviewees were asked to give their opinion about the AVs' safety and eco-friendliness, and to state whether or not they would be willing to use a self-driving vehicle. In addition to the responses to these questions, sociodemographic features were collected, including gender, age, level of education, and size of residence. Concurrently, the topics of the questions were analyzed by performing an in-depth literature review. Thereby, the results of the citizen survey could be compared to the latest state of expert knowledge.

G. Meyer and S. Beiker (Eds.): AVS 2019, LNMOB, pp. 78–87, 2020.
https://doi.org/10.1007/978-3-030-52840-9_8

2 Preliminary Considerations

In this study, only the highest levels of road vehicle automation were considered: SAE levels 4 and 5 in which no human driver is required. Furthermore, different scenarios of how automated driving could affect the future of mobility were identified and analyzed. These form the basis for the assessment and evaluation of the survey results:

Scenario 1: Driverless Vehicles Used Privately
In this scenario, there is no significant change in the current usage of vehicles. The only difference is that there is no human driver to perform the driving task. One person, household, or company will own the vehicle, which will be used for single trips and parked for the remaining time [13, 14, 20].

Scenario 2: Driverless Vehicles Complementing Public Transportation
In the second scenario, a driverless vehicle is rented just for the time of the ride [20]. The vehicle is no longer owned by a particular individual, household, or company. In this case, the trip is a service (mobility-as-a-service). It is possible that the vehicle is issued to one person only, or that it is shared with other travelers (ride sharing). Here, AVs are used in addition to the public transportation system, which remains like today [3, 14]. While the majority of the journeys will be covered by the public transportation systems, smaller AVs will serve to link areas to the greater transit systems.

Scenario 3: Driverless Vehicles Competing with Public Transportation
In the third scenario, the boundaries between motorized private mobility and public transportation are blurred. As travelling with AVs will become convenient and cheap, users may no longer have any particular motivation to use mass transportation systems. Instead, all trips will be made using self-driving, often shared, cars. The public (mass) transportation systems will be used less then today [4].

From an economic point of view, it is not very likely that the first scenario will come true. It is estimated that the cost of an AV will be 3,000–6,000 US$ higher than that of a conventional car [5]. Due to this higher up-front investment, it is more probable that people will prefer a short rental over purchase of an AV. A counterargument against the third scenario is that the possible positive aspects of self-driving vehicles will be undone if the usage of small vehicles for individual mobility increases, simply due to the high additional congestion caused by those cars. Thus, it is assumed that AVs will rather be used where they complement an efficient public transportation system, as described in scenario 2.

3 Results

In this paper, the results for the most relevant four of the twelve poll questions are presented[1]. For each of the topics, first, a diagram is shown that depicts the responses to the asked question. After that, the survey results are summarized, and any differences

[1] The full study (in German, [21]) is available for download from:www.iit-berlin.de/de/publikationen/automatisiertes-fahren-in-der-smart-city.

between the answers of male and female interviewees are highlighted. In addition, the responses of the different age groups (18 to 29-year-olds; 30 to 44-year-olds; 45 to 59-year-olds and 60 years or older ones) are examined briefly. Since no differences could be found for the sociodemographic features educational level and size of residential location, that data was not evaluated.

3.1 Potential of AVs for Quality of Urban Life

The first relevant survey question asked was: "How high do you assess the potential of self-driving vehicles to enhance the quality of life in cities, e.g. due to smoother traffic flow?".

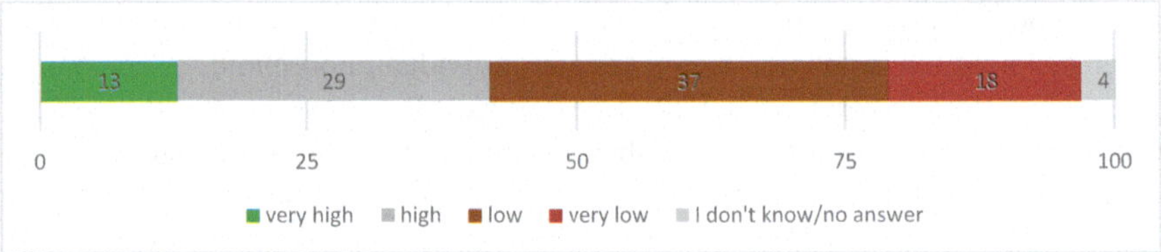

Fig. 1. Assessment of the potential of self-driving vehicles to enhance the quality of life in cities (The deviations from 100% are caused by rounding.)

According to the survey results (see Fig. 1), 42% of the interviewed persons think that the potential of self-driving vehicles to enhance the quality of life in cities is high (29%) or very high (13%). In contrast, 55% think that the potential is low (37%) or very low (18%). Females are noticeably more skeptical of whether AVs can enhance the quality of life in cities. 58% of the interviewed female persons think that the potential is (very) low, whereas 50% of the male respondents think the same. There are age-related differences as well: 63% of the 18 to 29-year-olds see a very high (29%) or high (34%) potential of AVs to enhance the quality of life in cities, but only 29% of persons aged 60 or older.

Current scientific and professional expert knowledge tells that AVs can improve the autonomy of people with reduced mobility. People with disabilities or limited mobility, as well as the elderly and children, may be able to travel more independently in AVs because as they may not be depending on an accompanying person anymore [9]. An AV could also be called and used independently if the necessary equipment is installed. Another positive aspect is that AVs can increase the safety of non-motorized road users. Since self-driving vehicles will obey the traffic rules and will keep an appropriate safety distance to other road users, e.g. cyclists or pedestrians, their subjective safety will be enhanced. This is possible because the current forms of motorized intimidation, driving too close, abrupt stops, and other user mishaps – will be avoided [1]. Furthermore, urban centers with high traffic and residential density could be relieved. Due to the higher driving comfort provided by AVs, people may be willing to live in more suburban areas instead of the city center [2]. Important for this would be that the higher mobility demand is served by an overall efficient transportation concept that is sustainable, and thus includes not just AVs but also emission-free vehicles and a high-performance public transport system. Additionally, AVs can cause a reduction of parking space. If scenario 2 comes true, the space needed for parked vehicles will be significantly lower than today. In Germany, only 9% of the current vehicle fleet is used at peak hours [17]. By reducing the amount of vehicles, the space that is currently occupied for parking will decrease. In addition, special automated valet parking garages may arise, in which an AV would not occupy more than one quarter of the space a conventional car would need. This is possible, because the space for the access roads and ramps, as well as the space around the car could be strongly reduced as security spaces can be removed, and no human (driver) needs to get in or out of the car. Besides, the ability to communicate with each other and with the infrastructure elements can lead to a smoother traffic flow. Thus, traffic-jams are less probable and the people will spend less time in cars for the same trips.

In summary, experts have identified several positive effects that AVs can have on the quality of life in cities. Apparently, the possible benefits are not obvious to the population, which in general underestimates the potential as shown by the survey. Since the citizens are the potential users, the public's awareness of the potentials of AVs should be raised. The aforementioned and the possibility to experience a functioning system – e.g. at field tests, would enhance their acceptance. If the AVs are better usable for citizens and the positive effects become visible, the acceptance can be expected to rise significantly.

3.2 Market Potential of AVs

The second most relevant question to the interviewees was: "Can you personally imagine to use a self-driving vehicle?".

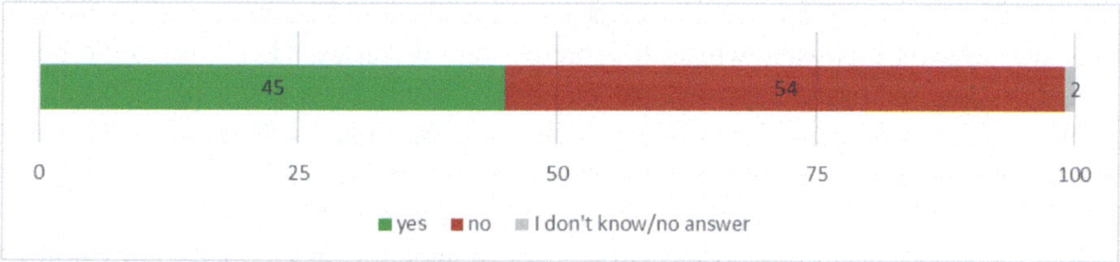

Fig. 2. Personal statement to the willingness to use self-driving vehicles (The deviations from 100% is caused by rounding.)

On this question, the survey found that 45% of the interviewed persons can imagine using a self-driving vehicle and 54% cannot (see Fig. 2). Males are specifically more willing to use an AV. 54% of the questioned male persons stated that they would use an AV, but only 36% of the females did. With increasing age the willingness to use a self-driving vehicle decreases. While 61% of the 18 to 29-year-olds can imagine using a self-driving vehicle, only 36% of the interviewed persons who are 60 years old and above can.

The assessment of the matter from the scientific and professional experts' perspective reveals that societal acceptance is an essential prerequisite if AVs are to be used ubiquitously. As stated in Sect. 2, the realization of the first scenario – where the ownership and use of AVs would be the same as with a conventional car today – is rather unrealistic. Market potentials will arise if economic advantages and safety gains become visible. If barriers of accessing and using the vehicles are kept small, acceptance of AVs may be expected to rise, and one of the scenarios 2 or 3 will come true. This implies that automobile manufacturers may need to adapt their business model, as sales volumes of automobiles may decrease and the manufacturers' role changes from sellers to service providers for e.g. car sharing. In this situation, the automobile manufacturers are lending a car for a limited period, and the users only pay for the ride. The automobile manufacturers are in charge of cleaning, maintaining, and refueling the vehicle. The current automobile ecosystem with OEMs, car sharing providers, taxi companies, and leasing or car rental companies would thus face significant change.

It should also be noted that many potential advantages of self-driving vehicles highly depend on their future usage: Besides the positive effects mentioned in Sect. 3.1, AVs would increase the capacity of the current roads: As AVs will be able to communicate amongst one another, the speed can be adapted according to the traffic situation, thus avoiding slow-moving traffic [7]. Also, vehicles have the highest fuel consumption when accelerating at a traffic light or in stop-and-go-traffic. By optimizing the driving behavior and reducing the acceleration and braking phases, the fuel consumption would thus decrease. Depending on the driving behavior of a human driver, AVs have the potential to reduce the fuel consumption by 15 to 30% [7].

At the same time, it is possible that AVs will travel in convoys. By communicating with other vehicles, the space needed for stopping can be reduced, as other vehicles can be informed about acceleration and braking actions in real-time. Thus, the reaction time in an AV platoon can be shorter – or rather equal to the latency – than in conventional vehicles. Velocity peaks when driving in convoys, and the effective driving time can be reduced [2]. As a consequence, the capacity of roads could be increased by 40% in cities and 80% on highways when all vehicles were driverless [12]. Due to the better traffic flow and increased capacity, the number of traffic jams could be reduced. An intelligent routing system, where the vehicles are distributed according to the roads capacity, would further contribute to this gain in capacity. Also, routes could be selected that are shorter or where fewer stops are necessary [7].

In conclusion, according to the survey, the population in Germany does not yet show a high affinity for the technology of self-driving vehicles. The ambivalent survey results reflect this. How AVs will be used in the future is currently an open question, but many optimization potentials are strongly dependent on that. A smoother traffic flow and fewer traffic jams are only possible if a substantial amount of the vehicles on the road are self-driving. It is also estimated that trips with AVs will be relatively cheap [10, 15]. At the same time, the current advantages of the public transportation systems – the possibility to read or to do something except for performing the driving task – would also applicable for trips in a private vehicle if it were self-driving. If all the trips would be performed with smaller, individual AVs, the possible positive effects will be undermined, though, and rebound effects will come to pass. As a result, the advantages of AVs would be reduced. To counteract that, the attractiveness of public transportation systems needs to be increased such that the main trips are performed with mass transportation systems, and only a few with AVs for individual transportation.

3.3 Privacy Issues of AVs

The third of the most relevant questions in the survey was: "Driverless vehicles can be the aim of hacker attacks. Do you agree?".

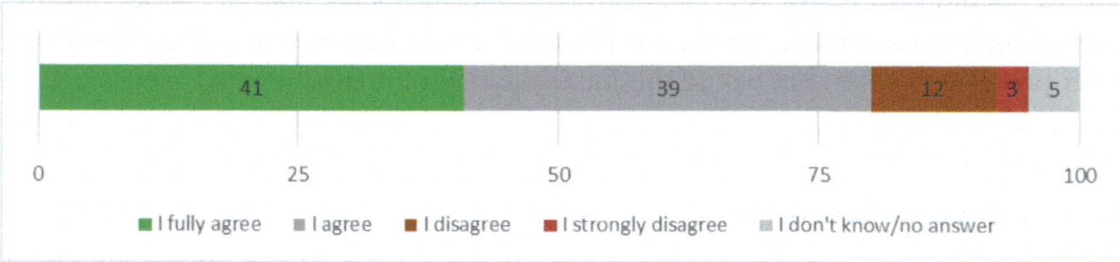

Fig. 3. Assessment of the statement, if self-driving vehicles can be the aim of hacker attacks

The survey results show that the population is seriously concerned when it comes to the cyber security of self-driving vehicles. 80% of the questioned persons agreed to the statement that driverless vehicles can be the aim of hacker attacks (see Fig. 3). Of these persons, 39% agreed and 41% agreed to the fullest. Only a small part (17%) of the interviewed persons does not think that AVs are open to hacker attacks. Interestingly, this is the only question for which the answers of the male respondents were more pessimistic than the female. 83% of the questioned males think that AVs can be attacked by hackers, but only 78% of the females do. Furthermore, it was surprising that the answers did not differ between the age brackets. Overall, that was the question with the highest consensus among the different groups.

Many expert discussions about AVs are focusing on the topics privacy and data security in particular. The AV's functionalities are built on cameras and other environmental sensors, controllers and data from the cloud to analyze traffic events, and to detect other road users. With a higher degree of automation, more interfaces to controllers are necessary, leading to a higher number of potential weak spots. Hackers would thereby be able to access sensitive personal information, driving data, or other connected devices like mobile phones. At the same time, the hacker attacks could also be targeted to manipulate the system and cause personal injuries.

From the populations' point of view, privacy is a very sensitive topic. The German Federal Government recently summoned an Ethics Commission that is supposed to give advice on whether automated driving is socially acceptable and desirable or not. One result was that sufficient data protection for all road users needed to be guaranteed the acceptance of AVs. Furthermore, it would be required that any person involved gives their consent before his or her data are released [6]. This means that the automobile manufacturers have to make great efforts to protect the privacy of the AVs and the road users they interact with. At the same time, the collected data can be an opportunity for the public authorities. The evaluation of traffic data of AVs – anonymized according to the data protection regulations – can help municipalities to control the traffic and optimize the infrastructure [8].

To conclude the findings on this question, cyber-attacks on self-driving vehicles can aim at interfering with the driving behavior, but also at intercepting personal data. The concerns of the population are widely shared by the scientific and professional expert community. Given the fact that hackers can assume the control of security-critical functions – specifically acceleration, steering, and the brakes – self-driving vehicles should be treated as critical infrastructure elements that are subject to high requirements and scrutiny. The higher IT security requirements would go along with higher production costs, but would minimize the risk of non-authorized accesses at the same time.

3.4 Road Safety Benefits of AVs

The fourth and final question cited from the survey, here, was: "The road safety in cities can be improved by using driverless vehicles. Do you agree?".

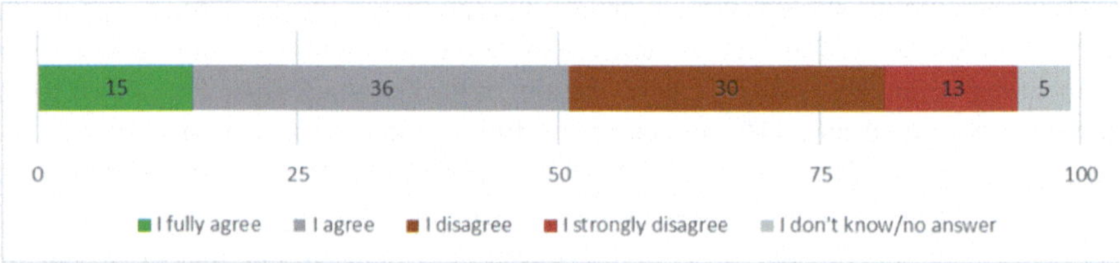

Fig. 4. Assessment of the statement, if the road safety in cities can improve by using driverless vehicles (The deviations from 100% is caused by rounding.)

According to the survey results (see Fig. 4), 51% of the interviewees think that the road traffic will become safer when AVs are used. 36% of those asked agree that the road safety in cities can be increased by using driverless vehicles, and 15% agree to the fullest. This question shows the highest discrepancy between women's and men's opinions. 62% of the interviewed male persons think that self-driving vehicles will make the traffic safer, whereas only 42% of the females agree with that statement. Noteworthy as well is that the consent to that statement is decreasing as age increases. Of the interviewed 18 to 29-year-olds still 73% agree with the statement, but only 42% of the persons who are 60 years old or above do.

According to the scientific and professional expert knowledge, the most important benefit of self-driving vehicles is that they will increase road safety. According to several studies, more than 90% of road accidents are caused by human error [18]. Because more and more driver assistance systems are used – e.g. anti-lock brake, electronic stability control, lane departure warnings adaptive cruise control, or driver fatigue detector – a reduction of the number of accidents is already noticeable. By using the emergency braking assistant, e.g., at least 20% of the passenger car crashes can be avoided [16]. Thus, it can be assumed that further automation will minimize human errors while driving and the road safety will increase.

In public debates of the decision-making criteria of AVs, it is rather the ethical aspects, such as the following questions that are much discussed: How will the self-driving vehicle 'decide', if it has the 'choice' to run into a child or an elderly person? Which moral decisions can be made by AVs? And, who is liable when an accident occurs? The Ethics Commission mentioned before was appointed to answer these questions. An essential outcome was that nobody, neither human nor machine, would be allowed to balance between lives in a dilemma situation. Hence, no rules or decision making processes can be programmed into the software. The same commission concluded that automobile manufacturers are liable for accidents that occur while driver assistance systems are active. Therefore, it has to be clear who is in charge while driving, the vehicle system or the driver. This information needed to be documented and saved at all times [6].

For the conclusion, it should be noted that the German Traffic Safety Association in 2007 stated the 'Vision Zero'. The aim is that there are no more fatalities or serious injuries in road traffic [11]. To achieve that objective, drastic measures still have to be taken, as there are still more than 3,000 road fatalities every year in Germany [19]. One option seems to be to use self-driving vehicles given their promises of higher safety. From that point of view, the use of AVs would be preferred. Though, it is important to note is that it still has to be proven that a driverless vehicle is safer than human drivers in a majority of the traffic situations, particularly in complex environments such as cities. Therefore, extensive real and virtual tests will have to be made to validate the system's performance. Only after that, the user acceptance will rise.

Of further interest is that even though more than half of the interviewed persons think that road safety can increase by using self-driving vehicles, only 44% can imagine using an AV. That allows the combined conclusion that an increase in safety is not necessarily a decisive factor for using AVs, at least not in view of the respondents to the poll. This is making clear again, that public awareness of pros and cons will be essential for a wide acceptance and deployment of AVs.

4 Conclusions

The automation of vehicles is going to change urban life tremendously in the future. It promises an accident-free traffic, optimized traffic flow, and inclusive mobility services. To understand how the German population is feeling about AVs, a representative survey was conducted, which was further compared to scientific and professional expert knowledge in a technical assessment. A central outcome was, that the population is underestimating the potentials of AVs compared to the current opinion in the scientific community. Furthermore, differences in the answers were found that are connected to sociodemographic features of the interviewed group, with females in general being more skeptical towards AVs and their potential. At the same time, those questioned between the ages of 18 to 29 were more open to the new technology, and further assessed the potential of AVs more positively. The older the interviewed persons were, the more critical they were. Other sociodemographic features like the educational level or whether the person lives in rural or urban environments did not correlate to their answers, though. The only question where less deviations between survey respondents and expert knowledge were visible is the question about cyber-security. In general, it is possible that the poll showed a general skepticism towards robotic technologies and not specifically towards AVs. To find that out, further research would be needed, while surely more efforts need to be placed in creating public awareness of the pros and cons of AVs.

It should be noted that the study was conducted in 2019 before the beginning of the COVID-19 crisis. During the pandemic, the worldwide mobility demand changed significantly, with a massive drop in passenger numbers in public transportation systems and shared cars. In sight of these circumstances, it is questionable, whether the survey results would still be the same, today. It is self-evident that mass transportation systems are used less, because it is hardly possible to keep the safety distance in a confined space. Furthermore, there are concerns about the cleanliness of vehicles. As an alternative, the usage of bikes and private cars are increasing. In many cities, roads have been transformed into pop-up pedestrian zones or bike lanes, such that those using the soft modes can keep distance from one another. During the pandemic, a scenario of individuals using small and shared self-driving vehicles instead of public transport may appear the most appealing. A reduction of the usage of car and ride sharing, as well as taxis and other mobility services is noticeable, though. An explanation might be that the population has doubts, whether the vehicles are disinfected properly. So, self-sanitizing capabilities might be a very relevant feature of AVs in the future. It should be considered, though, that overall mobility has shrunk by up to 40% temporarily during the pandemic, because people worked from home, schools and shops were closed, and private meetings were prohibited, all affecting any kind of transportation service. Hence, it is difficult to derive a clear change in the survey results from the behavior during the pandemic. While it is well imaginable that individual transportation in a shared AV could be more appealing now than before the pandemic, this conclusion cannot be made easily, as demand in shared vehicles has been dropping, too. Overall, it is likely that the skepticism towards the new technology and especially the concerns about the cyber security of AVs remain, and that any new features that appear beneficial in response to COVID-19 will not necessarily be met with enthusiasm only by the population. This insight that is clearly calling for a stronger involvement of citizens into the advancing AV design process applying co-creation approaches.

Acknowledgements. The authors gratefully acknowledge support by the Verein Deutscher Ingenieure (VDI) who initiated this study. They are also indebted to the members of the Institute for Innovation and Technology (iit) of VDI/VDE-IT in Berlin for their assistance, especially to Lorenz Hornbostel and Volker Wittpahl for managing the project, as well as to Doris Johnsen and Jörg Dubbert for their contributions to the state of expert knowledge assessments.

References

1. Alessandrini, A., Campagna, A., Site, P.D., Filippi, F., Persia, L.: Automated vehicles and the rethinking of mobility and cities. Transp. Res. Procedia **5**, 145–160 (2015)

2. Anderson, J., Kalra, N., Stanley, K., Sorensen, P., Samaras, C., Oluwatola, O.: Autonomous vehicle technology: a guide for policymakers. RAND Corporation (2016)
3. Arndt, D.-I. W.-H., Drews, F., Hertel, M., Langer, V., Wiedenhöft, E.: Integration of shared mobility approaches in Sustainable Urban Mobility Planning (2019)
4. Beiker, S.A.: Einführungsszenarien für höhergradig automatisierte Straßenfahrzeuge. In: Maurer, M., Gerdes, J.C., Lenz, B., Winner, H. (eds.) Autonomes Fahren, pp. 197–217. Springer, Heidelberg (2015). https://doi.org/10.1007/978-3-662-45854-9_10
5. Bernhart, W., et al.: Autonomous Driving. Disruptive innovation that promises to change the automotive industry as we know it - it's time for every player to think:act! (2014)
6. BMVI: Ethik-Kommission. Automatisiertes und vernetztes Fahren (2017)
7. Brown, A., Gonder, J., Repac, B.: An analysis of possible energy impacts of automated vehicle. In: Meyer, G., Beiker, S. (eds.) Road Vehicle Automation. LNM, pp. 137–153. Springer, Cham (2014). https://doi.org/10.1007/978-3-319-05990-7_13
8. Buchner, B.: Datengetriebene Geschäftsmodelle rund um das vernetzte Auto. Grundrechtsschutz im Smart Car. D, pp. 59–73. Springer, Wiesbaden (2019). https://doi.org/10.1007/978-3-658-26945-6_4
9. Cyganski, R.: Autonome Fahrzeuge und autonomes Fahren aus Sicht der Nachfragemodellierung. In: Maurer, M., Gerdes, J.C., Lenz, B., Winner, H. (eds.) Autonomes Fahren, pp. 241–263. Springer, Heidelberg (2015). https://doi.org/10.1007/978-3-662-45854-9_12
10. Deloitte: Urbane Mobilität und autonomes Fahren im Jahr 2035. Welche Veränderungen durch Robotaxis auf Automobilhersteller, Städte und Politik zurollen (2019)
11. DVR: Vorfahrt für "Vision Zero". Vision Zero. Keiner kommt um. Alle kommen an
12. Friedrich, B.: Verkehrliche Wirkung autonomer Fahrzeuge. In: Maurer, M., Gerdes, J.C., Lenz, B., Winner, H. (eds.) Autonomes Fahren, pp. 331–350. Springer, Heidelberg (2015). https://doi.org/10.1007/978-3-662-45854-9_16
13. Gruel, W., Stanford, J.M.: System effects of widespread use of fully automated vehicles—three scenarios. In: Meyer, G., Shaheen, S. (eds.) Disrupting Mobility. LNM, pp. 135–148. Springer, Cham (2017). https://doi.org/10.1007/978-3-319-51602-8_9
14. Heinrichs, D.: Autonomes fahren und stadtstruktur. In: Maurer, M., Gerdes, J.C., Lenz, B., Winner, H. (eds.) Autonomes Fahren, pp. 219–239. Springer, Heidelberg (2015). https://doi.org/10.1007/978-3-662-45854-9_11
15. Keeney, T.: Mobility-as-a-service: Why self-driving cars could change everything (2017)
16. Kühn, M.: Der Beitrag des automatisierten Fahrens zur Erhöhung der Verkehrssicherheit. Grundrechtsschutz im Smart Car. D, pp. 43–55. Springer, Wiesbaden (2019). https://doi.org/10.1007/978-3-658-26945-6_3
17. Nobis, C., Kuhnimhof, T.: Mobilität in Deutschland - MiD Ergebnisbericht. Studie von infas, DLR, IVT und infas 360 im Auftrag des Bundesministers für Verkehr und digitale Infrastruktur (FE-Nr. 70.904/15), Bonn, Berlin (2018)
18. Radke, T.: Elektrifiziert, automatisiert, vernetzt - herausforderungen und chancen für die antriebsentwicklung von morgen. VPC – Simulation und Test 2016. P, pp. 269–283. Springer, Wiesbaden (2017). https://doi.org/10.1007/978-3-658-16754-7_17
19. Unfallentwicklung auf deutschen Straßen 2017, Wiesbaden (2017)
20. Zmud, J., et al.: Research to examine behavioral responses to automated vehicles. In: Meyer, G., Beiker, S. (eds.) Road Vehicle Automation 5. LNM, pp. 53–67. Springer, Cham (2019). https://doi.org/10.1007/978-3-319-94896-6_5
21. Halil, Y., Hornbostel, L., Johnsen, D., Dubbert, J., Meyer, G., Wittpahl, V.: Automatisiertes Fahren in der Smart City. IIT, Berlin (2020)

Democratising Driverless Futures: Five Lessons for Public Dialogue on AVs

Jack Stilgoe(✉) and Nicole Badstuber

Department for Science and Technology Studies, University College London, 22 Gordon Square, London, UK
{j.stilgoe,n.badstuber}@ucl.ac.uk

Abstract. Self-driving cars (automated vehicles or AVs) are no longer just a laboratory experiment. In some parts of the world, prototypes are starting to appear on public roads. The thoughts of developers have understandably turned to their relationship with the members of the public who could become the users of the technology, stakeholders in its development or interested bystanders. The people involved in innovation are likely to have confidence in their technology and emphasise its potential benefits and its safety. Members of the public may see things very differently. With past technologies, the tendency has been to reject the views of members of the public as ill-informed or seek to change public attitudes. The evidence from previous controversies is that just talking without listening represents a bad approach. There is a need for ongoing public dialogue, not just top-down salesmanship. This is particularly important when a technology is being developed, as AVs are, not just in private laboratories, but in public.

1 Technologies in Public: From Deficit to Dialogue

1.1 Introduction

Towards the end of the 20th Century, genetically modified (GM) crops were a technology full of promise. Scientists were excited about the possibilities of more precise crop improvement, and companies saw clear economic opportunities. Alongside realistic proposals for incremental improvement ran hyped-up claims that the technology would benefit everyone, particularly the world's poorest people. In the US, the market for which most of the new varieties were developed, the technology was largely successful. Monsanto, the agricultural chemicals giant, was one of the first big players to spot the financial potential of developing new varieties of crop that could then be patented and sold to farmers. In Europe, some citizens and interest groups took issue with Monsanto's model and the claims they were making. The controversy that ensued took companies and scientists by surprise. From the 1990s onwards, supermarkets in the UK responded to a public backlash by refusing to stock genetically modified foods.

J. Stilgoe and N. Badstuber—The authors are part of *Driverless Futures?*, (driverless-futures.com) a three-year social science project, funded by the UK's Economic and Social Research Council, aiming to understand expert and public views on AVs and inform policymaking.

G. Meyer and S. Beiker (Eds.): AVS 2019, LNMOB, pp. 88–96, 2020.
https://doi.org/10.1007/978-3-030-52840-9_9

In France, where systems of agriculture are very different from those in the US, field trials of the technology came under huge public pressure. While there had been more than 1,000 trials in 1998, by 2004 there were fewer than 50, half of which were destroyed by protesters (Bonneuil, Joly and Marris 2008).

So while genetically modified foods have become a fact of life in the US, in much of Europe it is all but impossible to grow, sell or buy them. A public controversy meant that companies have missed out on markets, scientists have missed out on research opportunities and farmers and consumers have missed out on new innovations. Until the GM crops controversy, the dominant way in which scientists and policymakers thought about the public and public trust was to presume that if citizens knew more about science and innovation, they would support scientists' and innovators' vision of the future. This approach has been labelled the 'deficit model' (Wynne 1991). The diagnosis was that members of the public lacked scientific understanding, which explained their scepticism of new technologies. The suggested solution was to teach them science. In both the US and Europe in the 1980s and 90s, scientific bodies created programmes for science communication and the 'public understanding of science'.

The problems with the assumption that to know science was to love it quickly became apparent. With technologies like nuclear power that were in some places extremely controversial, the more citizens learnt, the more worried some became. People, often the most educated people, were unwilling to just accept the answers that scientists were offering. They had their own questions. Policymakers began to realise that not only was there a need for genuine public dialogue, but also a need to have this dialogue early, while the technology was still in development (Wilsdon and Willis 2004). Our project – *Driverless Futures*? – is trying to put this idea into action, bringing public views and social science to a debate that has until now been dominated by technological questions. The history of the public debate around genetically modified crops can help inform the choices society might make about self-driving cars. In simple terms, we should pay attention to five big lessons.[1]

1.2 Lesson 1: Debates About New Technologies are Never just About Science and Technology

The first generation of genetically modified crops suggested a profound disruption to people's everyday lives. The people developing the technology were understandably excited about the benefits for consumers, farmers and food producers. But these groups and other citizens saw the technology in the context of their own lives and wanted to ask about what it meant for the future of farming, trade, the economy, the rural landscape and more.

Food is something everyone can relate to, as is mobility. When non-experts talk about AVs they should therefore not be expected just to stick to technological questions. They may be excited about the possibilities, but they will also want to ask how

[1] The social research on public attitudes to GM crops is wide-ranging, but an important early example is (Grove-White et al. 1997). A summary of insights from public dialogue on GM is in this report: https://sciencewise.org.uk/wp-content/uploads/2018/12/Talking-about-GM-published.pdf.

their commutes, their families' lives and the places in which they live could change in a world in which AVs are normal. Some will also wonder, as people did with GM foods, who really benefits from the technology. It is vital to remember that the general public are more than just a pool of potential consumers.

1.3 Lesson 2: People are Citizens as Well as Consumers

Social research with members of the public on attitudes to GM crops revealed not just individual concerns, but collective ones. People were worried about what futures were being created with new technology and whether future worlds were ones in which they wanted to live. When it comes to transport, which has obvious planning implications, people are used to being consulted. If AVs are going to change the world, people will want to have a say, and they will not expect the market alone to realise the opportunities they see or resolve their concerns. They will want to discuss who could benefit, who could lose out, who should be in charge, and who will take responsibility. One only needs to compare the vastly different transport systems in the world's cities to anticipate that one-size-fits-all AV approaches are likely to elicit very different public responses in different places.

As with other parts of public life, citizens won't just engage as voters, nor only when they are asked their opinion. Some groups may feel that their interests are served or threatened by AVs. Some NGOs have already taken strong positions even while the technology remains uncertain. Many others have not yet worked out their relationships with AVs, but the range of possible issues is likely to be wide.

1.4 Lesson 3: It's About More Than Safety

With GM foods, the developers of the technology assumed that public concerns would be dominated by questions of risk: Will they be safe to eat? Where environmental NGOs were involved, developers presumed that questions would centre on environmental risk. In either case, these questions were seen as scientific ones, and therefore negotiable in scientific terms. Public misunderstandings of the science were used as a pretext for rejecting public concerns. Developers were therefore surprised by the expansion of public controversy into areas of politics.

The novelty of GM foods meant that members of the public did have concerns about regulation for safety. People wondered how, if the technology was so new, we could be sure of its safety. Here, GM food developers were stuck in what Steve Rayner (2004) called a 'novelty trap': the benefits were advertised as radically exciting but the risks were seen as incremental and straightforward. (The vociferous argument about the labelling of GM foods was a product of this novelty trap).

Members of the public were also concerned that new interventions in complex ecosystems challenged existing scientific understanding. The perceived 'unnaturalness' of GM technologies is one common manifestation of such concerns. Finally, the GM controversy raised issues of economic inequality. People and NGOs were concerned about intellectual property, patenting and the livelihoods of developing country farmers. Given these concerns, people lacked confidence in in the ability of scientists, companies and governments to understand and regulate full set of concerns about new technology.

At the height of the controversy, around 1999, there was a damaging collapse in public trust. A survey of European citizens found that only 6% named universities as institutions they trusted to tell the truth about genetically modified (GM) crops, while 26% of people named environmental NGOs (Haerlin and Parr 1999).

1.5 Lesson 4: People in Power Need to Listen as Well as Talk

The developers of new technology are understandably enthusiastic about what they have to offer, but innovation cannot be a monologue. Innovation is a conversation between needs and possibilities. For AVs, there is a need to understand what people's real hopes and fears are. The uncertainties are huge. Companies and policymakers are at the moment making some assumptions about safety, but we still have no idea how safe is safe enough. The acceptability of risk is not something that can be decided by engineers. Do people think being safer than a human driver on average is acceptable? The social science of risk perception would suggest not. Levels of acceptable risk can vary by orders of magnitude even among different transport modes. People tend to magnify risks that they feel are catastrophic, out of their control, new and with little connection to benefit (Slovic 1987). But the point is that we don't yet know. We don't know whether people will have concerns about who owns AV data. We don't know how people will balance values like privacy against convenience. We don't know what people think about the interpretability of machine learning. We don't know whether it matters to people if this is public transport or private, personal or shared. We don't know how all of these things will vary from place to place. So we need to listen. But the conversation can't end there. If innovators are going to ask people what they think, they need to say how they are going to change direction in response. Otherwise it is public engagement for engagement's sake.

1.6 Lesson 5: Be Clear on Why You are Doing Public Engagement

When it comes to public engagement, there is often more emphasis on the *how* than the *why*. Organisations need to be clear on why they want to tell the public something or ask for people's opinions. If it's to sell a particular technology, or to lobby for policy change, be honest about that. People will see right through it if not. Is it to persuade or is it to empower? Is it to open up the debate to new perspectives or to close it down? We can talk about the broad motivations for public participation being normative (it's the right thing to do), instrumental (it helps us do what we want to do, by creating more trust) or substantive (it gives us social intelligence that helps us make better decisions or better technologies) (Fiorino 1990). Often, organisations have instrumental motivations. They think that being seen to reach out will convince the public they are trustworthy. And often, if this is the aim, the activity backfires. Good engagement costs money and time. It's only worth an organisation paying for it if they are going to learn something. A useful reflective question is whether public engagement is intended to open up debates or to close them down (Stirling 2008).

In the UK, the controversy over GM crops convinced policymakers and scientists of the need for new forms of institutionalised public dialogue. In 2004, the UK Government created a new organisation, Sciencewise, to commission and organise deliberative exercises on issues of new science and technology. Over the last 15 years, Sciencewise has supported more than 50 dialogue exercises on topics ranging from data science to decarbonisation, from flooding to the future of cities.

2 Public Insights on Driverless Futures

In 2018, Sciencewise and the UK Centre for Connected and Autonomous Vehicles commissioned a public dialogue exercise that brought together more than 150 members of the public in five parts of Britain to discuss, over three days, the possibilities and concerns of AVs. In three locations, participants were able to experience AV technologies, riding in, variously, a low-speed driverless shuttle, a prototype self-driving car and a simulator. At the end of the third day, having heard from experts, studied information about AVs, conducted their own homework and discussed among themselves, the participants provided recommendations for policy.

Summarising such lengthy and wide-ranging is challenging. The first thing to note is that public opinion is, as with other new technologies, ambivalent. People are both excited and worried. Individuals may, without contradiction, be both enthusiastic users of a technology while also being sceptical its development and its place in society. We only need to think of our relationships to our smartphones to interpret this ambivalence as normal, but it is often seen as paradoxical or problematic for policy (Kearnes and Wynne 2007). For AVs, there was real enthusiasm among the Sciencewise participants for the potential of the technology, particularly to bring mobility to people who for reasons of disability, location or income, lacked good transport options. But there was also a real concern that these benefits might not be realised if the technology was developed and managed badly.

The full report of the dialogue exercise highlights a few specific issues that provide a constructive challenge to dominant ways of thinking and talking about AVs. Participants were sceptical of the idealised vision of the driverless future offered by AV developers. To pick one example, Waymo promises to: "improve the world's access to mobility while saving thousands of lives now lost to traffic crashes" without requiring changes to public infrastructure. People wondered how realistic such claims were. They thought that the technology would arrive first in the places where the infrastructure, road conditions and potential markets were already set up. Other places and people would therefore lose out. They also saw the technology's potential to worsen problems, such as increased congestion, while alleviating others. These participant thought the technology would realistically only come to urban or suburban places:

> "I live in a rural area, so I can't see those pods impacting me, I would still need a car."

> "Is there a need for it in a village? If they don't have it, they'll be stuck."

> "So what you're saying is that people in the countryside can't get one of your motors [AVs]? That's a bit unfair isn't it?"

Others questioned who would really benefit and who would foot the bill:

> Participant A: "Infrastructure has been my biggest issue."
> Facilitator: "Will the infrastructure need to change?"
>
> Participant B: "It'll have to."
> Participant A: "Significantly."
> Facilitator: "Who should pay for the infrastructure?"
>
> Participant A: "Users should pay. I don't think taxpayers should pay."

Some participants thought that access to the technology was likely to highly unequal, and were concerned about injustice:

> "They are also putting a lot of money into this technology, so I don't think they are going to be spending money on public transport. So I think there will be effects. So we need a guarantee that going forward it's not going to deplete [public transport investment] further."

Concerns about inequality also extended to questions about employment for professional drivers. Participants wondered…

> "What about people who are employed to drive?"

> "How can we avoid people losing their jobs?"

> "If I was a lorry driver I wouldn't be happy"

People also challenged the conventional narrative of AV safety benefits by asking how we would know if the technology itself was safe. There was a recognition that testing would need to happen in public, but this would not be straightforward. As one person put it,

> "There will be risks, we will learn from accidents, but I don't want my family to be those on the back of which the learning happens."

The participants wanted to see independent systems for understanding, improving and verifying the safety of new AV technologies. They were concerned not just about the risks of crashes, but also the risks of system failure, both malicious and unintentional, and (particularly for women) the personal security dangers of travelling without a driver, potentially on vehicles shared with others.

While AVs are often framed as offering additional mobility options, much of the discussion reflected a concern about a loss of freedom and control. Some of this was about giving up control of a vehicle to a car, but the bigger issue of people's control over how they got around was a more important recurring theme. One participant, reflecting on conversations they had had after an earlier session, was surprised that, in reacting to AVs, people expressed "a sense of losing their independence, not having their car in the garage and being able to use it when they wanted." The paradox was that greater autonomy of vehicles was frequently perceived as a loss of autonomy for humans, as suggested in the following quotations:

> "I wouldn't want to go completely with it. I would still want to be able to take control of the car [...]. I wouldn't want to abandon my car altogether."

> "Cars were liberating for the working classes and older people. This seems to be restricting choice."

> "What if someone doesn't want a driverless car?"

> "Why would I agree to a system that restricts my choice to go where I want when I want, and with my dog?"

> "It will be for the greater good, but it worries me. I don't know if I personally can make all the changes required to adapt to this world."

The final concern, about how the changes individuals may need to make to their own lives, should not be downplayed. Most of the participants recognised that technologies do not just plug-and-play; they work in particular conditions and often the conditions need to change in order for them to work. For AVs, participants quickly realised that the issue went beyond just who or what was doing the driving. Questions of individual and shared mobility and public and private transport quickly became relevant. The participants' recommendations therefore focussed on the conditions they would like to see for the responsible development of the technology. In summary, people felt that they would be more comfortable…

- If the technology is proven to be safe and secure
- If the benefits of the technology are widely available
- If the technology is good for society and jobs
- If we're in control of our transport
- If there is clear guidance on accountability
- If new regulatory bodies are created[2].

Given the potential benefits and concerns, most participants were reassured that the dialogue exercise had been initiated by Government, suggesting that companies would not be left to their own devices. As one of the more enthusiastic participants said in a video recorded after the dialogues,

> "I didn't know that we are not left alone. It's not left alone in big corporations' hands. Government knows about this and they're doing something, and I'm really, really happy about that".[3]

Taken together, these attitudes present a substantial challenge to developers, most of whom are currently focussed on the technical challenge of safely navigating complex environments and the economic challenge of identifying profitable business models for

[2] These principles and other conclusions are in the CAV public acceptability dialogue Engagement report, Sciencewise, 24 July 2019, available here: https://assets.publishing.service.gov.uk/government/uploads/system/uploads/attachment_data/file/837958/cav-public-acceptability-dialogue-engagement-report.pdf.

[3] CAV Public Acceptability Dialogue – Video of workshop participants, 7 Feb 2020, https://www.youtube.com/watch?v=_BKm0o16ofA.

products and services. But public opinion need not and, if the lessons of GM crops are to be heeded, *should* not be seen as a barrier. Instead, it can be a resource to empower good policy and good system design.

2.1 Governing AVs

Public dialogue can be a useful prompt for policy and innovation, but it should not just be seen as a one-off exercise. Policymakers, companies and others should ask themselves how to make open, deliberative public engagement a normal part of their activities. The regulation of new technologies can take many forms. We do not yet know whether an AV will be regulated like private cars, like transport systems, like data-driven technologies or like something else. We could see a standardisation of international approaches, or national governments could take very different approaches to technologies that seem otherwise identical.

By the time of the genetically modified crops controversy in Europe, much of the regulatory apparatus in both the US and Europe had already been constructed. US policymakers had, in the 1980s, chosen to regulate the *products* of genetic modification, asking whether the need strains of crop were substantially equivalent to their conventionally bred counterparts. European regulators had adopted a more precautionary approach to the process of genetic modification, emphasising novelty and inviting additional scrutiny (Pollack 2010).

With AVs, many of the rules that will govern their development are not yet written. Given the uncertainties, open dialogue is important. In this mode, developers' one-way communication of information remains important. It will be vital to tell members of the public what is happening and why. But developers should not expect members of the public to trust them just because they are told to. Responsible companies can help explain the limits of technology as well as the potential benefits. Clarity about expectations will be important at a time when confusion suits the companies that are most prone to hype. Building a healthy conversation around AVs is not easy, but the alternative could be bad for developers, bad for governments and bad for the public.

References

Bonneuil, C., Joly, P.-B., Marris, C.: Disentrenching experiment: the construction of GM—crop field trials as a social problem. Sci. Technol. Hum. Values **33**(2), 201–229 (2008). https://doi.org/10.1177/0162243907311263

Fiorino, D.: Citizen participation and environmental risk: a survey of institutional mechanisms. Sci. Technol. Hum. Values **15**(2), 226–243 (1990)

Grove-White, R.M.: Uncertain World: Genetically Modified Organisms, Food and Public Attitudes in Britain. CSEC and Unilever, London (1997)

Haerlin, B., Parr, D.: How to restore public trust in science. Nature **400**(6744), 499 (1999). https://doi.org/10.1038/22867

Kearnes, M.B., Wynne, B.: On nanotechnology and ambivalence: the politics of enthusiasm. Nanoethics **1**(2), 131–142 (2007). https://doi.org/10.1007/s11569-007-0014-7

Stirling, A.: "Opening up" and "closing down" power, participation, and pluralism in the social appraisal of technology. Sci. Technol. Hum. Values **33**(2), 262–294 (2008)

Pollack, M.: When Cooperation Fails: The International Law and Politics of Genetically Modified Foods. Oxford University Press, Oxford (2010)
Rayner, S.: The novelty trap: why does institutional learning about new technologies seem so difficult? Ind. High. Educ. **18**(6), 349–355 (2004). https://doi.org/10.5367/0000000042683601
Slovic, P.: Perception of risk. Science **236**(4799), 280–285 (1987). https://doi.org/10.1126/science.3563507
Wilsdon, J., Willis, R.: See-through science: why public engagement needs to move upstream (2004). http://www.demos.co.uk/publications/paddlingupstream. Accessed 28 Feb 2020
Wynne, B.: Knowledges in Context. Sci. Technol. Hum. Values (1991). https://doi.org/10.1177/016224399101600108. https://journals.sagepub.com/doi/10.1177/016224399101600108

Automated Vehicles & Vulnerable Road Users: Representing the Under-Represented

Justin M. Owens[1(✉)], Michael Clamann[2], David Aylor[3], Stacy A. Balk[4], Jana Lynott[5], Maya Pindeus[6], Amy Rosepiler[7], Lauren Silverstein[8], and Francis Gemperle[8]

[1] Virginia Tech Transportation Institute, 3500 Transportation Research Plaza, Blacksburg, VA 24061, USA
jowens@vtti.vt.edu

[2] University of North Carolina Highway Safety Research Center, 730 Martin Luther King Jr. Blvd, Suite 300, Chapel Hill, NC 27599-3430, USA
clamann@hsrc.unc.edu

[3] Insurance Institute for Highway Safety, 4121 Wilson Boulevard, 6th Floor Arlington, VA 22203, USA
daylor@iihs.org

[4] Human Factors/Engineering Integration Division, Office of Vehicle Crash Avoidance and Electronic Controls, National Highway Traffic Safety Administration, 1200 New Jersey Avenue SE, Washington DC 20590, USA
Stacy.Balk@dot.gov

[5] AARP Public Policy Institute, 601 E Street NW, Washington DC 20049, USA
JLynott@aarp.org

[6] Humanising Autonomy, Somerset House, Strand, London WC2R 1LA, UK
maya@humanisingautonomy.org, Maya@humanisingautonomy.com

[7] Burgess & Niple, Inc., 5085 Reed Road, Columbus, OH 43220, USA
amy.rosepiler@burgessniple.com

[8] Uber ATG, 50 33rd Street, Pittsburgh, PA 15201, USA
lauren.silverstein@uber.com

Abstract. This chapter provides an overview and recap of the AVS 2019 Breakout Session *AVs & Vulnerable Road Users: Representing the Under-Represented*, including summaries of research and perspective presentations from leading experts in the field and needs identified through discussion among panelists and the session audience. The session identified a range of necessary actions and research needs including defining technological, improving education about automation and advanced technology, and using these to build public trust.

Keywords: Equity · Automation · Vulnerable road user · Pedestrian · Cyclist · Disability · Mobility · Accessible transportation · Aging · Older

G. Meyer and S. Beiker (Eds.): AVS 2019, LNMOB, pp. 97–107, 2020.
https://doi.org/10.1007/978-3-030-52840-9_10

1 Introduction

Emerging technologies including automated vehicles (AVs), connected infrastructure, and shared mobility systems are transforming and disrupting the field of transportation. In many instances, these technologies offer the potential for improved safety, increased access and mobility across different populations, and/or enhancement of existing transportation services (such as first/last-mile access to transit). At the same time, these technologies may create new concerns or barriers to travel for members of vulnerable groups including pedestrians, bicyclists, people with disabilities (PWD), and children, collectively referred to as Vulnerable Road Users (VRUs). These concerns may be related to safety (e.g., finding ways for AVs and VRUs to effectively share the roadway), human factors (e.g., ensuring that children PWD are able to engage with new technology), or equity (e.g., people without smartphones may be put at a disadvantage).

VRU safety has been a persistent concern across the automotive age. While vehicle driver and occupant deaths have tended to decrease over the past several decades in both absolute numbers and as a function of vehicle miles traveled [1], pedestrian, bicyclist, and motorcyclist deaths have remained relatively steady or increased over the past 40 years. There are numerous potential failure points that could lead to conflicts between human drivers and VRUs. These include detection issues (e.g., a driver overdriving his/her headlights at night or in other situations with poor visibility), poor vehicle control or inability to react in time to avoid a collision due to impairment or drowsiness, or slowed physiological responses due to age or disability. While technology has continuously advanced to improve within-vehicle safety, these critical points of human driver limitation have remained relatively unaltered for decades; in some cases, advances in driver comfort technology such as multimedia entertainment and traditional cruise control could even reduce a driver's ability to properly recognize and avoid VRUs.

The advent of automation brings the potential for new safety models of driver interaction with VRUs, including for the first time systems designed to assist drivers with the failure points that could directly affect VRU safety. Features like pedestrian detection, forward collision warning, blind spot warning, and automatic braking, which used to be the province of luxury vehicles, are in production across vehicle fleets. The promise of the convergence of these features in the coming years into fully automated driving systems provides yet more opportunity for improvements in VRU safety. However, new technology often comes with new challenges, and it is crucial to recognize that automation will not be an instant panacea for current risks faced by VRUs; in fact, automation will present new risks and barriers for VRU safety.

The goal of this breakout session was to expand the discussion surrounding the interactions among AVs and VRUs, with particular focus this year on the safety of under-represented groups including children and seniors. To accomplish this, the breakout session was structured to enable attendees to accomplish several distinct goals:

- Learn about the latest advances in technology and research designed to facilitate AV/VRU interaction
- Hear new and varied perspectives about the importance of under-represented groups, planning and design in ensuring safe and fair mobility in future transportation systems
- Gain a better understanding of current VRU transportation needs and how these may be moderated by advancing technology
- Engage with researchers, experts, and safety advocates to exchange ideas and identify research needs
- Work collaboratively to develop a list of interdisciplinary research needs statements that support vulnerable road users

After an introduction and overview, the first portion of the session included a series of talks that presented recent research on ways that VRUs, including children, seniors, pedestrians and cyclists, are accounted for and considered in cutting-edge transportation system development, including advanced sensor and communication technology, predictive modeling, and roadway design. After research presentations, the session transitioned to small group breakouts, in which the presenters worked with groups of attendees to discuss research needs and areas for advancement. The session lasted for a single afternoon from 1:30–5:30. The following two sections of this chapter provide overviews and summaries of the information presented and research needs developed during the two portions of the breakout session.

2 Research and Perspective Presentations

During the first section of the session, experts in a range of domains related to VRU safety provided a few opening remarks detailing their recent research in the field and/or perspectives on the issues surrounding VRUs and automation/connectivity. After these remarks, a moderated panel discussion was conducted to allow experts to discuss topics of interest amongst themselves and with the audience. In this section we summarize the presented remarks.

David Aylor of the Insurance Institute for Highway safety provided an overview of the effectiveness of currently available pedestrian detection technology. Stacy Balk, formerly of Leidos, described how VRU safety could be improved in the future through implementations of connected vehicle technology. Michael Clamann of UNC then discussed the unique challenges of deploying AVs in school zones, and Jana Lynott or AARP explained the need to have comprehensive inclusion of transportation services through AVs. Maya Pindeus, CEO of Humanising Autonomy, described advanced technologies for detecting and predicting VRU intent. Amy Rosepiler, representing Burgess & Niple, Inc. discussed the potential impact of AVs on infrastructure, and the possible related effects on VRUs. Finally, Lauren Silverstein and Francine Gemperle of Uber ATG provided an overview of how they are developing a model of verbal and nonverbal road sharing interactions with VRUs

Making intersections safer with advanced pedestrian detection technology
David Aylor, Manager of Active Safety Testing, Insurance Institute for Highway Safety

Pedestrian motor vehicle crash deaths have increased 53% since reaching their low point in 2009, and account for 17% of all crash fatalities. In 2018, more than 6,000 pedestrians were killed and approximately 75,000 pedestrians were injured in vehicle crashes on public roads in the United States. The increase in fatalities has occurred mostly in urban areas, away from intersections, on busy roads, and at night. Crashes also increasingly involve SUVs.

Crash avoidance technology that is able to detect pedestrians can reduce the death toll. These systems use sensors to monitor in front of the vehicle and warn drivers of potential collisions with pedestrians. Most of the systems can automatically apply the brakes when a crash is imminent. The Highway Data Loss Institute studied insurance claim rates for Subaru models equipped with and without an optional pedestrian detection system. The study found pedestrian injury claims were 35% lower among vehicles with pedestrian detection.

The Insurance Institute for Highway Safety (IIHS) began rating such systems in 2019. IIHS evaluates vehicles in three scenarios: an adult stepping into the street in the path of the oncoming vehicle with an unobstructed view, a child darting into the street from behind two parked cars, and an adult standing near the side of the road in the travel lane, facing away from traffic. Vehicles are assigned one of four ratings—no credit, basic, advanced or superior—based on their ability to mitigate or avoid a collision with the pedestrian dummy. Superior-rated vehicles reduce their speed dramatically in all three tests and in most cases avoid hitting the pedestrian dummy, eliminating or greatly reducing the risk of severe injury. Advanced-rated systems achieve significant speed reductions in most of the tests. Vehicles that earn a basic rating or receive no credit fail to slow the vehicle significantly in one or more of the tests.

Considering the role of connectivity in future pedestrian and bicyclist safety

Stacy Balk, Human Factors Program Manager (former), Leidos

The percentage of pedestrian and bicyclist fatalities due to vehicle collisions has risen steadily over the past decade. Factors known to increase risk of crashes include environmental conditions (weather, lighting and road surface), infrastructure (road geometry, grade, crowded urban settings, and traffic control), driver behavior (avoidance maneuvers and speed), and road user characteristics (driver/pedestrian impairment and distraction). Proven solutions such as road diets, crossing refuges, improved signal timing, and reduced speed limits are effective, but not adequate in reducing the overall percentage of VRU fatalities. However, connected vehicle technologies presents the opportunity to potentially reduce this gap in VRU safety through the implementation of new pedestrian safety applications.

To develop a better understanding of how connectivity can improve VRU safety, our research established a tested bed for emerging V2P technologies at Turner-Fairbank Highway Research Center (TFHRC). Using this test bed, we evaluated 3 forms of V2P systems.

- System 1: Camera-based aftermarket safety device
- System 2: OEM camera and radar based detection system
- System 3: Smartphone based pedestrian-to-infrastructure application

System 1 (camera based aftermarket safety device) was found to have reliable detection of both pedestrians and bicyclists when traveling in a straight line. However, horizontal and vertical curves and weather had a strong influence on the system's ability to detect pedestrians and bicyclists. System 2 (OEM camera and radar based detection system) provided reliable detection and alerts of pedestrians, but was less reliable in detecting bicyclists. System 2 was also subject to the same negative influence from roadway geometry and environmental conditions as system 1. System 3 (Smartphone based pedestrian-to-infrastructure application) addresses most of the limitations apparent in camera and radar based systems, but required all pedestrians to use a smartphone and for a data/server infrastructure to be developed.

One of the limitations with conducting of V2P system evaluations is the lack of mature, market-ready, and publicly accessible products. To overcome this limitation, we utilized virtual reality (VR) to conduct human factors research. The first human factors study conducted at the TFHRC Virtual Reality Lab (VRL) was an evaluation of alert modalities for a Vehicle-to-Bicycle Collision warning system. The alert modalities evaluated were haptic, visual, and audible. In this study, participants were asked to navigate through an urban virtual environment using the VRL's bicycle simulator. The participants were subject to a variety of hazards, such as vehicles running red lights and vehicles overtaking the cyclists. We then evaluated alert modality effectiveness using metrics such as evasive maneuver and reaction time. This connected collision warning system was still in development during the time of data collection and the VR bicycle simulator afforded us the opportunity to conduct human factors testing before the technology was fully developed.

Automated vehicles and schools: An analysis of deployment issues

Michael Clamann, Senior Human Factors Engineer, UNC Highway Safety Research Center

Automated vehicles face substantial challenges accurately and reliably detecting and recognizing pedestrians, who are more difficult to identify, predict, and protect in the event of a crash compared with other road users. The challenge of predicting their behaviors is amplified and further complicated every weekday around school zones where large numbers of vehicles interact with children entering and departing schools, each of which is a unique dynamic transportation environment with its own approach to regulating traffic. It is critical that drivers follow local rules to ensure the safety of children entering and leaving the school zone, because nearly every student walks on school property to some extent, whether traveling from a bus, a car, or from home. Currently, students driving or being driven to school generates 10% of vehicle trips in the morning (7:00 to 9:00am) and eight percent of vehicle miles traveled [2]. Therefore, if the goal is to improve safety through broad deployment, automated vehicles will eventually need to be designed to operate in and around school zones.

The combination of a dynamic environment and critical safety requirements will be a challenge for automated vehicles, which often do not perform reliably when faced with uncertainty. Their deployment will undoubtedly require further innovation along with changes to local policies governing travel in school zones. Automated vehicles will require updated infrastructure, regulation and technology that should be carefully evaluated by developers and school stakeholders to determine their feasibility. Safe deployment of automated vehicles around schools will require a coordinated effort between ADS developers, local planners, school administrators, and other community members. With these issues in mind, researchers from the Pedestrian and Bicycle Information Center (PBIC) identified nine recommendations for stakeholders when preparing for the deployment of automated vehicles around schools [3]:

1. Developers should ensure pedestrian detection systems account for children
2. Traffic safety educators should account for ADS deployment in future materials intended for children and adults
3. School administrators should plan to update local pick-up and drop-off procedures to account for ADS-specific regulations and capabilities
4. Developers should work with school transportation stakeholders to identify low-cost solutions that support safe ADS navigation on school property
5. ADS developers and entities who develop training programs for crossing guards should develop and validate procedures for crossing guards
6. ADS must be able to detect when they enter and exit school zones and prioritize posted speed restrictions
7. Localities should ensure roadway infrastructure is maintained to facilitate accurate detection by ADS sensors
8. ADS should consistently comply with school zone traffic regulations
9. ADS test plans must account for school zones

It is important to note that the engineering, infrastructure, education, and enforcement issues in these recommendations are inherently intertwined. The solution is likely going to be a compromise and open communication between local transportation programs, their affected communities, and companies developing automated vehicles to consider all the interconnected elements of the transportation system, from street, to sidewalk, to building, with considerations for stakeholders of all ages. It is likely that there will not be a one-size-fits-all solution that will apply to every jurisdiction. It is important that stakeholders are included in conversations about future deployment.

Universal Mobility as a Service: the equitable way to roll out AV technology

Jana Lynott, AICP, Senior Strategic Policy Advisor—Transportation & Livable Communities, AARP Public Policy Institute

Tremendous innovation is happening in the transportation sector today, but with too little thought being given to the needs of non-drivers of advanced age and others with disabilities. Take, for instance, Google Transit, or any number of trip planning apps that now provide real time bus and rail information at the click of a button on one's smart phone. New services like Lyft and Uber have already shown us the possibilities of on-demand transportation. And literally overnight those of us in DC and many other cities across the globe woke up to find thousands of new dockless bikes and scooters on our sidewalks and available to any of us with a smart phone and credit card.

Many promoters of autonomous vehicles tout the technology as the solution to immobility for older adults and people with disabilities who do not drive. While technology advancement will increase access for many, to assume that the technology alone is a panacea is extremely misguided. Numerous considerations exist for enabling mobility for the older and disability population, including:

- Many people need that extra assistance in/out of the vehicle or may not be able to cognitively navigate even the most intuitive apps.
- We will always need the human element available through a call-center, as trip navigators, or simply to offer assistance and security on board the vehicle.
- AV technology offers no guarantee that we will have any more wheelchair accessible vehicles on our roads than we have today.
- LIDAR, infrared and other cameras are expensive, and will be for the foreseeable future. If we simply replace our human-driven vehicles with driverless vehicles, and change nothing else, we will only push more people out of affordable transportation.

AVs are part of the solution if rolled out right. If not, they could make the situation even more difficult. Autonomous vehicle technology will be most effective at addressing the systemic problem of inequity in access if it rolls out through Universal Mobility as a Service. Universal MaaS builds upon basic industry concepts of Mobility as a Service to ensure that all users' needs are met, regardless of income, geographic location, disability, or age. *Universal* refers to both a comprehensive inclusion of all available transportation services within a mobility platform and accessible design throughout the system. Specifically, Universal MaaS includes the following:

- Fully integrating specialized transportation within the MaaS framework
- Applying the principles of universal design to all facets of the system
- Fostering shared-use mobility, rather than overreliance on car ownership and solo driving (or riding in the case of autonomous vehicles)
- Offering customers, including those with special needs, a variety of transportation options, along with seamless payment system, and traveler information
- Integrating fixed-route and demand responsive public transportation, as well as taxi service, ridehailing, carpooling, carsharing, bikesharing, and volunteer transportation.

One of the prerequisites of realizing Universal Mobility as a Service is for human services transportation providers to modernize and embrace the widescale adoption of a common digital language that will enable providers to share trip data and provide more seamless service. FlexDanmark, in IT company managed by Denmark's five regional transportation authorities, uses the Standardiserat Utbyte av Trafik Information (SUTI) standard to share trip data across more than 500 unique demand responsive transportation providers. It is a world-class model for coordinated transportation services.

The disruption we are engaged in offers a window of unprecedented opportunity when we as planning and aging professionals, policymakers, business leaders, and advocates have a choice. We can recreate – or perhaps exacerbate the seemingly intractable problems of our current system. Or we can harness technological change to innovate and provide a safe, convenient and reliable transportation system for each and every one of us, no matter our stage in life, physical abilities or mental capacity.

For more information visit www.aarp.org/futureoftransportation.

Human behaviour prediction for accident and near miss prevention
Maya Pindeus, COE, Humanising Autonomy

About: Humanising Autonomy (HA) has developed real-time behaviour analysis and intent prediction software for accident and near-miss prevention, increasing the safety of VRUs such as pedestrians and cyclists. The technology understands the full range of human behaviours, and predicts numerous intents across different environments using only visual cameras. Initially developed for autonomous vehicles, the software integrates with all levels of autonomy, including driver-assist features for human-driven vehicles such as Automated Emergency Brakes, front and near-side driver alerts, or rear-mirror camera monitoring systems.

To support this software, we are building a large-scale, comprehensive VRU dataset across multiple cities, with which it can build models to understand the full range of human behaviour. It has built a state-of-the-art object detection and tracking software, purpose built for VRUs in the urban environment to meet the dynamic challenges of the automotive industry today. By going beyond the bounding box, our technology can improve VRU safety in urban environments today and help global partners reach Vision Zero.
Key outputs of this technology include:

1. Intent prediction for accident and near miss prevention (risk prediction 2 s faster than human response)
2. False positive prevention to enable automated urban driving
3. Real time analysis and prediction of events around the vehicle
4. Customizable features that are adaptable to specific application requirements, environments, and cultures.

Key benefits of the technology include:

1. Accident reduction with pedestrians and vulnerable road users through integrated software.
2. Integrates with the vehicles vision sensors and feeds real time prediction into Path planning, Automated
3. Emergency braking or Driver alert
4. Quicker deployment of automated vehicles in cities.
5. Reduction of insurance premiums through improved driving safety.
6. Improved efficiency of operations for drivers in highly pedestrianized environments

I know what I know: Applying design to put pedestrians first
Amy Rosepiler, Urban & Complete Streets Project Manager & Director, Roadway Design, Burgess & Niple, Inc.

With the introduction of automated vehicle technology, the potential impacts to roadway infrastructure and, specifically, pedestrian related infrastructure are unclear. It is anticipated that AVs will need minimal infrastructure cues that are common today, such as signalized intersections, to seamlessly navigate corridors and communicate between vehicles and the environment around the vehicle. However, it is imperative to remember the vulnerable users in the corridor. Pedestrians are constantly interacting with their surroundings and are conditioned, taught or trained to behave a certain way (whether following cues from the surrounding infrastructure, or by making up their own rules). Pedestrians do expect or require certain infrastructure, especially to address ADA needs, and anticipate specific responses from the infrastructure.

As an engineer, it is imperative to ask, "How do I ensure my design considers the highest safety for the most vulnerable user? How do I loop in Towards Zero goals and make sure the pedestrian is placed first, regardless of the vehicle and driver (computer or human)?" While AV are anticipated to improve safety, the physics of a potential crash – in particular, a multi-ton vehicle striking a pedestrian – have not changed. As our society moves into the smart technology future it is important to remember we can design infrastructure to reinforce safety and meet the expectations of pedestrians even if the vehicle technology is not dependent upon that infrastructure. This design may need to be flexible to support AV technology, but should not lose focus on the priority in the corridor – the vulnerable road user. We can also continue to support this design with the safety data that has been collected and published thru the Highway Safety Manual and captured in crash modification factors (CMF) to ensure our designs are establishing a priority for the most vulnerable road.

Contextualizing Pedestrians: Beyond the Crosswalk

Lauren Silverstein & Francine Gemperle, Uber Advanced Technologies Group

Introduction

Our work is primarily focused on the planning that goes into creating a Self-Driving Vehicle capable of navigating complex environments. We study the behavior of Vulnerable Road Users (VRUs) to understand and model how people share the road so that we can turn that understanding into something the vehicle's autonomy system can participate in.

Road Sharing

Sharing the road with various transportation modalities is not a new concept; it has been a challenge for humans since the invention of the wheel. Each transportation innovation has forced humans to re-determine how to share space while in motion. Uber Advanced Technologies Group (ATG) aims to share the road safely with all road users.

Scene Complexity

In the workshop, we together looked at several real-life scenes by watching short videos of people and vehicles sharing the road. Understanding a scene means looking at the actors (people, cyclists, etc.), context (time of day, ped density), and environment (intersection, infrastructure, etc.). This breakdown is a powerful tool for understanding scene complexity and building models of how to share and engage.

Model of Interactions

We drafted a model of interactions - based on the contents of ISO TR23049 - to establish a framework for how people and vehicles can (and do) communicate with each other through different kinds of signals in the shared road space. There are highly consistent types of signals. The meaning of this signal is clear and is used regularly by different transportation modalities like cars, bikes, and even self-driving cars. There are the signals whose means are less certain and incredibly dependent on human concepts and cultures, such as waving, smiling, and gesturing. There is a rich set of communications, both verbal and nonverbal, on the road that are constantly in use. Roadsharing is not unique to Uber ATG - as a community we all need to grow this understanding of how to safely share the road.

3 Discussion and Action Items/Research Needs

Following the perspective presentations of each panelist, all workshop participants were prompted to engage in small group discussions to identify the most important issues related to automated vehicles and vulnerable road users. The discussion topic was deliberately broad to encourage a variety of participant perspectives. After brainstorming on issues in small groups, participants were then asked to come up with recommended steps to prepare for broad automated vehicle deployment. The results of this second exercise were shared with all workshop participants and used to prompt a discussion among all participants to collectively identify a set of research needs.

There was general agreement among the participants that the overarching question surrounding automated vehicles is how to develop a broad public understanding of when the technology is sufficiently safe for deployment. This combines the important (and often debated) question of "how safe is safe enough," as it relates to deploying automated vehicle technology, with effectively educating the public about how "safe enough" will be measured, and accurately communicating when it occurs. This overarching question included several research needs, which were categorized as being related to *technology readiness* or *public education*.

The first research need was to *define incremental readiness levels and measures of success for different operational design domains*. Rather than identify one broad definition for safety, the group determined that safety should be driven by context, and technology deployment readiness should be defined separately for different ODDs. The other idea related to technology readiness was to *understand the tradeoff between operations/mobility and safety*. At their extremes, safety can be assured if vehicle operations stop, and operations will be highly effective if we stop considering public safety. Of course, the actual tradeoff is somewhere in-between, and it is crucial to identify the compromise that properly weighs the needs of all types of VRUs.

On the topic of public education, participants expressed a need to *educate all road users about automated vehicle topics*. If road users, including VRUs, are going to interact with automated vehicles, it is important for everyone to understand their strengths and limitations. It is also important to understand how to effectively *maintain open communication among stakeholders*. Numerous populations will be affected by automated vehicle deployment, including those trying to improve their mobility options, VRUs interacting with moving vehicles, and law enforcement and emergency medical personnel, just to name a few. Automated vehicle developers should understand these various perspectives and incorporate them in their development efforts. There was also an interest in identifying *to what extent VRU behaviors can be influenced* with education and enforcement. There is a long history of using education and enforcement to influence road user behaviors, but additional work is required to determine how these tools can be applied effectively to automated vehicle deployment.

Furthermore, VRUs should be engaged in the ongoing *dialogue about how safe is safe enough* to make sure safety includes road users outside and inside the vehicles.

Finally, public trust is essential to the successful deployment of automated vehicles. Therefore, it is also important to ensure that automated vehicle *safety benefits help build user trust*. In the context of VRUs, this means that AVs should be designed to protect all types of road users outside the vehicle, independent of ability, background and socioeconomic status, and those benefits should be communicated in a way that ensures those groups will trust their abilities during and following broad deployment.

In conclusion, this session provided an opportunity for attendees with expertise in transportation and automation technology to engage in discussion with experts in industry, technology, planning, and mobility for vulnerable road users. We identified a series of valuable next research steps that will meaningfully contribute to the safe and successful advancement of automation to improve safety and mobility for vulnerable road users.

References

1. National Safety Council: Historical Fatality Trends (2018). https://injuryfacts.nsc.org/motor-vehicle/historical-fatality-trends/deaths-and-rates/
2. Kontou, E., McDonald, N., Brookshire, K., Pullen-Seufert, N., LaJeunesse, S.: U.S. active school travel in 2017: prevalence and correlates. Prev. Med. Rep. (2019). https://doi.org/10.1016/j.pmedr.2019.101024
3. Clamann, M.: Automated vehicles and schools: An analysis of deployment issues. In: Stanton, N., et al. (eds.) 10th International Conference on Applied Human Factors and Ergonomics and the Affiliated Conferences. Springer (2019)

Part IV Vehicle Systems and Technology Development

New Simulation Tools for Training and Testing Automated Vehicles

Jiaqi Ma[1], Chris Schwarz[2(✉)], Ziran Wang[3], Maria Elli[4], German Ros[4], and Yiheng Feng[5]

[1] University of Cincinatti, 765 Baldwin Hall, Cincinatti, OH, USA
jiaqi.ma@uc.edu
[2] National Advanced Driving Simulator, 2401 Oakdale Blvd, Iowa City, USA
chris-schwarz@uiowa.edu
[3] Toyota Motor North America, Mountain View, CA, USA
zwang050@ucr.edu
[4] Intel, Portland, OR, USA
{Maria.elli,German.ros}@intel.com
[5] University of Michigan, Ann Arbor, MI, USA
yhfeng@umich.edu

Abstract. Simulation offers the potential benefit of testing many miles in a variety of situations and environments. Not only can virtual testing environments be run quickly and in parallel, but a greater focus can be brought to bear on rare edge cases that need to be understood by ADS. They support simulation of sensor suites, environmental conditions, full control of all static and dynamic actors, maps generation and much more that enable automated vehicle simulations. They have large and growing communities that can contribute to the simulation ecosystem and develop use cases. This chapter presents ADS simulation research and capabilities discussed at the breakout session entitled "New Simulation tools for Training and Testing Automated Vehicles" at the 2019 Automated Vehicle Symposium in Orlando, FL. The section reviews key highlights and conclusions from three studies presented at the breakout session: 1) Responsibility Safety Sense (RSS) and software-in-the-loop (SiL) simulation; 2) augmented-reality-based testing with accelerated scenario design; and 3) human-in-the-loop testing for freeway cooperative merge.

Keywords: Simulation · Automated driving systems · Responsibility Safety Sense (RSS) · Augmented reality · Human in the loop

1 Introduction

The central problem of testing automated vehicles, or automated driving system (ADS), is broadly understood. The Rand corporation studied the problem with an analysis of reliability rates and concluded that hundreds of millions of miles, sometimes hundreds of billions, would have to be tested to ensure that ADS failure rates were better than human-driven vehicles [1]. Wachenfeld and Winner also used a statistical approach to evaluate ADS testing requirements [2]. Using a Poisson model for occurrences of accidents on German highways, the authors describe the so-called approval trap. Even

G. Meyer and S. Beiker (Eds.): AVS 2019, LNMOB, pp. 111–119, 2020.
https://doi.org/10.1007/978-3-030-52840-9_11

if an automated vehicle is safe, there is no good way to prove its safety. An even more sober assessment was given by Mobileye, in which the authors assume that ADS failures must be three orders of magnitude rarer than current rates to be accepted by society [3].

Simulation offers the potential benefit of testing many miles in a variety of situations and environments. Not only can virtual testing environments be run quickly and in parallel, but a greater focus can be brought to bear on rare edge cases that need to be understood by an ADS [4, 5]. Waymo reached a landmark 10 million miles of on-road driving with their ADS technology; however, they have driven 10 billion miles in simulation [6].

The availability of low-cost tools such as Unity3D and Unreal has inspired several efforts to construct ADS simulations, most notably CARLA [7] using Unreal and Airsim [8] which is available in both Unreal and Unity3D. An example list of open-source ADS simulation tools is shown in Table 1. Moreover, SaaS solutions have sprung up to assist ADS developers with their testing needs while hiding the complexity of integrating many different software tools and libraries. Metamoto and Cognata are examples of cloud-based SaaS products [9].

Table 1. An example list of open-source ADS simulation tools

Name	Developer	Website
CARLA	Intel	http://carla.org/
LGSVL	LG	https://www.lgsvlsimulator.com/
AirSim	Miscfrosoft	https://microsoft.github.io/AirSim/
DeepDrive	Voyage	https://deepdrive.voyage.auto/
Udacity SDC	Udacity	https://github.com/udacity/self-driving-car-sim

In contrast to simulation approaches for traditional vehicles, automated vehicle simulations must consider the environment to a much greater degree. The simulation of ADS sensors, including LiDAR, radar, cameras, and others, will be limited in fidelity by the level of realism offered in the virtual environment. Specifically, material types need to be encoded to allow distinctions among objects that are metal, concrete, biological, heat sources, etc. [9–12].

Simulation-based testing is not a panacea. Every component that is replaced by a virtual model raises the question of model validation. It may not be clear whether the problem of validating all parts of a simulation is feasible to meet testing requirements [2]. Therefore, simulation must be considered as one tool in the toolbox along with test track and on-road testing [10, 12]. In fact, novel hybrid testing strategies may combine test track or on-road testing with simulation in new ways [13–15].

This chapter presents ADS simulation research and capabilities discussed at the breakout session entitled “New Simulation tools for Training and Testing Automated Vehicles” at the 2019 Automated Vehicle Symposium in Orlando, FL. Speakers from Intel and Metamoto presented on simulation tools, respectively Responsibility Safety Sense (RSS) and software-in-the-loop (SiL) simulation. Yiheng Feng and Ziran Wang offered presentations on recent research conducted at Toyota Motors North America and the University of Michigan Transportation Research Institute.

2 Application of Responsibility Safety Sense (RSS) in CARLA Open Source Simulator

CARLA is a free, open-source automated driving simulator that aims to accelerate the introduction of new autonomous driving technologies [7]. CARLA has been designed from day one to support the development and validation of autonomous driving systems. The CARLA platform defines a versatile simulation API that gives users and developers control over all the elements of the simulation. In this way, it can be used for a wide range of applications, from sensor placement to the prototyping and testing of your planning and control algorithms (see Fig. 1).

Fig. 1. CARLA scenario runner helps to set up tests

A key design feature of CARLA is its scalable architecture, following a server-multi-client approach to allow for the distribution of computation into multiple nodes. The server takes care of running sensor simulation and updating the physics of the environment. Clients are developed using the CARLA API and they take care of a variety of tasks, such as controlling the traffic and running the Autonomous Driving stack under evaluation.

Through the API, clients can subscribe to different events, including sensor updates (e.g. LIDAR, cameras, radars, etc.), ground truth events (e.g., semantic segmentation messages, object annotations, etc.), and infractions, to name a few. These messages enable users to train and test new perception models, work on new planning algorithms and test new vehicle controllers. The platform also offers the tools required to create new traffic situations and maps.

Additionally, CARLA is integrated with an open-source implementation of the Responsibility Sensitive Safety (RSS) model published by Mobileye in 2017 for advancing the research, development, and verification of AV safety [3]. RSS formalizes human notions of safe driving, using a set of mathematical formulas and common-sense rules that are transparent and verifiable. The goal of RSS is to allow the AV to drive carefully enough so that it will not be the cause of an accident, and cautiously enough so that it can compensate for the mistakes of others. RSS within CARLA operates as a separate layer from artificial intelligence-based planning algorithms. It has the role of a "safety checker" that deterministically analyzes if the decisions made by the planning module conform to the model or not and proposes a proper response in cases where the ego vehicle is in a dangerous situation. RSS integrated with CARLA enables safety research and testing that can be verified without millions of miles of driving. Because RSS is a formal mathematical model, it can be proven correct, significantly reducing the validation burden.

Thanks to the modular and open philosophy driving the CARLA project and the aforementioned features, this simulation platform has become one of the default choices for the research, development, and testing of autonomous vehicles and other problems within the context of advanced transportation.

3 Unity-Based AV Simulation with V2X Communication and Human-in-the-Loop Integration

As a commercial game engine, Unity provides a suitable platform for developers to model AV simulation by integrating various sensing technologies. Additionally, AV models built in Unity can also be integrated with vehicle-to-everything (V2X) communication, as well as human-in-the-loop (HiL) simulation. In this section, a V2X-based advisory speed assistance (ASA) system is introduced in Unity, where pure AV model and HiL mode can be switched based on the existence of driver inputs [16].

3.1 Objective

An ASA system is designed in Unity as a head-up display (HUD), where the driver of AV can decide whether to control the vehicle speed or not. If no, then the advisory speed will be directly executed by the low-level vehicle controller (actuator) and the vehicle is considered as an AV. If yes, then HiL simulation can be conducted and the driver will execute the advisory speed by stepping on the accelerate and brake pedals.

V2X communication enables information sharing among various vehicles, so the advisory speed shown on the HUD considers the surrounding traffic environment of the ego vehicle. Such systems improve the safety, mobility, and environmental performances compared with no V2X communication is involved.

3.2 Advisory Speed Assistant

ASA is designed for the driver of an AV (or semi-automated vehicle), which is enabled with V2X communication. The main purpose of this system is to provide auxiliary information to suggest the vehicle longitudinal speed for the driver to follow, so the vehicle can be driven in a safer and more efficient way. The recommended vehicle longitudinal speed can be calculated by the proposed longitudinal motion controllers and shown to the driver through HUD.

As shown in Fig. 2, the HUD design is developed by the "Canvas" in Unity, which is the area that contains all advisory elements. The "Canvas" is a game object with a canvas component on it, and all advisory elements such as the text shown on the HUD, are children of such a canvas. The canvas is shown as a rectangle in the scene view of Unity, allowing us to easily position our HUD design on the windshield and right above the dashboard. Since we attach the game object canvas as a child of the game object vehicle, the HUD design becomes the grandchild of the vehicle and is fixed on that position on the windshield.

Fig. 2. Design of HUD-based ASA in Unity

Additionally, the target vehicle of this ego vehicle is identified through V2X communication and the on-board camera, and it is marked with a "TARGET LEADER" sign on top of its roof by HUD, notifying the driver which vehicle to follow after merging into the mainline. Note that we also design the side mirrors and the rear-view mirror in the vehicle cabin, allowing the driver to observe vehicles running behind while conducting the merging behaviors.

3.3 Human-in-the-Loop Simulation

Once the advisory speed is computed and displayed to the driver by the aforementioned HUD design, the driver of the controlled ego vehicle needs to execute it to Unity in the longitudinal direction, meanwhile also keeps the ego vehicle at the center of the current lane in the lateral direction. A vehicle controller is modeled to transfer driver lateral and longitudinal inputs into vehicle dynamics based on the game object "WheelCollider" in Unity. Generally speaking, a "collider" in Unity defines the shape of an object for the purposes of physical collisions, while the "WheelCollider" is a special "collider" designed for vehicles with wheels in Unity. It has built-in collision detection, wheel physics, and a slip-based tire friction model.

$u_x \in [-1, 1]$ is acted as the steering input of the vehicle, which allows two front wheels to steer around the local y-axis in Unity. The steer angle θ can be calculated by

$$\theta = \theta_{max} \cdot u_x \tag{1}$$

where θ_{max} is the user-defined maximum steering angle of the front wheels. The steer angle θ can then be transmitted to the front wheels to steer the vehicle by the "WheelColliders[i].steerAngle = θ" command in Unity's C# scripting API, where $i = 0, 1$ denotes the front left and right wheels. $u_y \in [-1, 1]$ is implemented using the accelerate or brake inputs of the vehicle, which allows the vehicle to move along the local y-axis by accelerate/brake torque, generated by its wheels. When the driver steps on the accelerator pedal, $u_y \in (0, 1]$. Alternatively, $u_y \in [-1, 0)$ when the driver steps on the brake pedal.

The thrust τ_{thrust} and brake torque τ_{brake} can be calculated by

$$\tau_{thrust} = \frac{\tau_{tmax}}{n_{wheels}} \cdot u_y, if u_y \in [0, 1] \tag{2}$$

$$\tau_{brake} = \tau_{bmax} \cdot u_y, if u_y \in [-1, 0) \tag{3}$$

where τ_{tmax} denotes the full thrust torque of the vehicle over all wheels, τ_{bmax} denotes the full brake torque of each wheel, $n_{wheels} = 2$ if the vehicle is either front-wheel drive or rear-wheel drive, and $n_{wheels} = 4$ if the vehicle is four-wheel drive.

The thrust torque τ_{thrust} is then be transmitted to the wheels to accelerate the vehicle by the "WheelColliders [i].motorTorque = τ_{thrust}" command in Unity, where $i = 0, 1, 2, 3$ denotes the front left, front right, rear left and rear right wheels, respectively. Similarly, the brake torque τ_{brake} can be transmitted to each wheel to decelerate the vehicle by the "WheelColliders [i].brakeTorque = τ_{brake}" command in Unity's scripting API. With aforementioned control, the driver can execute the advisory speed illustrated on the HUD and conduct cooperative maneuvers with other vehicles.

4 Testing and Evaluation of Autonomous Vehicle with Naturalistic Driving Data and Augmented Reality

Closed test facilities, which can test real CAVs in a controlled environment, has its unique advantages over simulation and pub-road testing. Meanwhile, due to lacking realistic background traffic, the number of traffic scenarios that can be tested in closed test facilities is limited. Moreover, how to systematically generate testing scenarios remains a big challenge. In this study, a new testing evaluation framework is proposed to address these two limitations. The augmented reality technology is used to generate virtual background vehicles to interact with real test CAVs, while a testing scenario library generation (TSLG) framework is proposed to generate a set of critical scenarios for a given operational design domain (ODD), based on naturalistic driving data (NDD) analysis and surrogate model (SM) simulation. The proposed framework is implemented at the Mcity test facility at the University of Michigan with a Level 4 CAV.

The motivation of the AR testing environment is to generate background traffic to affiliate CAV testing and evaluation since most of the testing scenarios require interactions with other vehicles. The AR environment combines a real-world test facility and a simulation platform together. Movements of test CAVs in the real world is synchronized with simulation and data of simulated vehicles are fed back to test CAVs. Test CAVs can interact with simulated traffic as if in a realistic traffic environment. Compared to using real vehicles, simulated vehicles can be easily controlled in generating different scenarios with reduced cost, safety concerns and higher accuracy. For instance, when the test CAV fails in a safety-related test and hits a red light running vehicle, no one will actually be hurt. Moreover, such tests can be executed repeatedly to increase test efficiency. The AR environment can serve as a pre-step before involving real vehicles to ensure algorithms are thoroughly examined and parameters are fine-tuned. The proposed system is very beneficial to assessing CAV technologies in a cost-effective fashion. More details of the AR environment can be found in [17].

With generated background virtual vehicles, the next step is to design the maneuvers of the vehicles to generate testing scenarios. Because the number of total scenarios is huge even an ODD is given, a systematic way to identify critical scenarios, which have higher values in testing, is necessary (i.e., TSLG). Toward this end, first, we propose a new definition of scenario criticality as the combination of maneuver challenge and exposure frequency. The maneuver challenge measures the dangerous or difficulty level of a scenario, while the exposure frequency denotes the probability of the scenario occurring on the public roads. The maneuver challenge is quantified by the surrogate model (SM) simulation. A human driving model is used as the SM to represent general features of CAV behaviors because human drivers are natural baselines. The exposure frequency of a specific scenario is estimated from the Safety Pilot Model Deployment (SPMD) dataset [18] collected by the University of Michigan Transportation Research Institute (UMTRI). Finally, an optimization-based searching method is developed to find critical scenarios (i.e., scenarios with higher criticality values) and constructs the library [19].

The proposed framework is implemented at Mcity with a cut-in case study. A level 4 CAV is utilized in the test [20]. A virtual cut-in vehicle is generated in the AR environment and performs cut-in with different range and range rate settings. Different combinations of range and range rate are sampled from the constructed library with ε-greedy policy to balance the exploration and exploitation of the library. Field test results show that the proposed method can converge to the same performance index (e.g., crash rate) as public road testing. Meanwhile, the number of required tests reduces significantly by 9.87×10^4 times, which greatly accelerates the evaluation process. For more details about the case study, readers can refer to [21].

5 Conclusions

This chapter presents the lecture notes on new simulation tools for ADS testing and analysis. Simulation is now fundamental to the continued development of automated driving systems. They are based on open source or freely available software, such as Unity and Unreal game engines. They support simulation of sensor suites, environmental conditions, full control of all static and dynamic actors, maps generation and much more that enable automated vehicle simulations. They have large and growing communities that can contribute to the simulation ecosystem and develop use cases.

Using these tools and developing supporting testing theories and standards are still in their infancy. Such discussions need to be focused on the sharing of testing experiences and methodologies to enhance testing capability and efficiency. Future explorations are recommended on the following perspectives:

- Provide information on new AV simulation tool product and the latest research
- Learn how researchers and engineers are using such tools in enhancing their own work
- Discuss how the tools can be adapted to various types of research, e.g., integrate into a human-in-the-loop simulator, hardware-in-the-loop simulator, or macroscopic simulation
- Collect ideas for research needs.

References

1. Kalra, N., Paddock, S.M.: Driving to safety: how many miles of driving would it take to demonstrate autonomous vehicle reliability? Transp. Res. Part A Policy Pract. **94**, 182–193 (2016)
2. Wachenfeld, W., Winner, H.: The new role of road testing for the safety validation of automated vehicles. In: Automated Driving, pp. 419–435. Springer (2017)
3. Shalev-Shwartz, S., Shammah, S., Shashua, A.: On a formal model of safe and scalable self-driving cars. arXiv preprint arXiv:1708.06374 (2017)
4. O'Kelly, M., Sinha, A., Namkoong, H., Tedrake, R., Duchi, J.C.: Scalable end-to-end autonomous vehicle testing via rare-event simulation. In: Advances in Neural Information Processing Systems, pp. 9827–9838 (2018)
5. Zhao, D., Huang, X., Peng, H., Lam, H., LeBlanc, D.J.: Accelerated evaluation of automated vehicles in car-following maneuvers. IEEE Trans. Intell. Transp. Syst. **19**(3), 733–744 (2017)
6. Etherington, D.: Waymo has now driven 10 billion autonomous miles in simulation (2019). https://techcrunch.com/2019/07/10/waymo-has-now-driven-10-billion-autonomous-miles-in-simulation/
7. Dosovitskiy, A., Ros, G., Codevilla, F., Lopez, A., Koltun, V.: CARLA: an open urban driving simulator. arXiv preprint arXiv:1711.03938 (2017)
8. Shah, S., Dey, D., Lovett, C., Kapoor, A.: Airsim: high-fidelity visual and physical simulation for autonomous vehicles. In: Field and Service Robotics, pp. 621–635. Springer (2018)
9. Leitner, A., Holzinger, J., Schneider, H., Paulweber, M., Marko, N.: Seamless tool chain for the verification, validation and homologation of automated driving. In: Validation and Verification of Automated Systems, pp. 165–176. Springer (2020)
10. Koopman, P., Wagner, M.: Toward a framework for highly automated vehicle safety validation. SAE Technical Paper0148-7191 (2018)
11. Fadaie, J.: The State of Modeling, Simulation, and Data Utilization within Industry: An Autonomous Vehicles Perspective. arXiv preprint arXiv:1910.06075 (2019)
12. Wood, M., et al.: Safety first for automated driving. In: Aptiv, Audi, BMW, Baidu, Continental Teves, Daimler, FCA, HERE, Infineon Technologies, Intel, Volkswagen 2019. https://www.aptiv.com/docs/default-source/white-papers/safety-first-for-automated-driving-aptiv-white-paper.pdf
13. Feng, Y., Yu, C., Xu, S., Liu, H.X., Peng, H.: An augmented reality environment for connected and automated vehicle testing and evaluation. In: 2018 IEEE Intelligent Vehicles Symposium (IV), pp. 1549–1554. IEEE (2018)

14. Liu, H., Feng, Y.: Development of an augmented reality environment for connected and automated vehicle testing. In: CCAT Project No. 2, University of Michigan, Ann Arbor, Transportation Research Institute2019-UMTR-4, 2019. http://hdl.handle.net/2027.42/149453
15. Wang, C., Winner, H.: Overcoming challenges of validation automated driving and identification of critical scenarios. In: 2019 IEEE Intelligent Transportation Systems Conference (ITSC). pp. 2639–2644: IEEE (2019)
16. Wang, Z., Wu, G., Boriboonsomsin, K., Barth, M.: Cooperative ramp merging system: agent-based modeling and simulation using game engine, SAE Int. J. CAV, **2**(2) (2019)
17. Feng Y., Yu, C., Xu, S., Liu, H.X., Peng, H.: An augmented reality environment for connected and automated vehicle testing and evaluation. In: Presented at 29th IEEE intelligent vehicle symposium, Changshu, China (2018)
18. Sayer, J.R., Bogard, S.E., Buonarosa, M.L., LeBlanc, D.J., Funkhouser, D.S., Bao, S., Blankespoor, A.D., Winkler, C.B.: Integrated vehicle-based safety systems light-vehicle field operational test key findings report (2011)
19. Feng, S., Feng, Y., Yu, C., Zhang, Y., Liu, H.X.: Testing scenario library generation for connected and automated vehicles, Part I: methodology. arXiv preprint arXiv:1905.03419 (2019)
20. Xu, S., Peng, H., Song, Z., Chen, K., Tang, Y.: Design and test of speed tracking control for the self-driving lincoln MKZ platform. IEEE Trans. Intell. Veh. (2019)
21. Feng, S., Feng, Y., Sun, H., Bao, S., Misra, A., Zhang, Y., Liu, H.X.: Testing scenario library generation for connected and automated vehicles, Part II: case studies. arXiv preprint arXiv:1905.03428 (2019)

Human, Machine, Sensor, Infrastructure: All Together Against Cyberattacks in AV

Jonathan Petit[1(✉)] and Victor Murray[2]

[1] Qualcomm Technologies Inc., 90 Central Street, Boxborough, MA 01719, USA
petit@qti.qualcomm.com
[2] Southwest Research Institute, 6220 Culebra Road, San Antonio, TX 78238-5166, USA
victor.murray@swri.org

Abstract. The Automated Vehicle's ecosystem relies on four main components: Human, Machine, Sensor, and Infrastructure. Each component is designed with its own threat model and requirements in mind. However, to ensure end-to-end security and resilience against cyberattacks, we should consider the interoperability of each security model. In this chapter, we give an overview of the state-of-the-art of security in Human Factors, Machine Learning, Sensors, and Infrastructure, before highlighting the research challenges to solve in order to design end-to-end resilience in AVs.

Keywords: Automated vehicle · Security · Machine Learning · Human factor · Sensor

1 Introduction

Automobile complexity is increasing at an amazing rate due to expanding connectivity, automation, and the desire to improve the driver's experience. As complexity increases, so do the vulnerabilities and the need to harden automotive systems against cyber-attacks. This chapter reviews cybersecurity impacts on automated vehicles with a focus on:

- Human Factors - Automated vehicles will need to interact with humans both inside and outside the car, and human factors must be considered for advanced levels of autonomy.
- Machine - Machine learning holds great promise with its ability to accurately classify objects, among other things, but these algorithms have been deceived by researchers, demonstrating a vulnerability in their perception.
- Sensors – Are used to track the vehicle and surrounding objects and are the inputs to control system installed on automated vehicles. However, they are limited in their ability to detect and are susceptible to external attacks such as spoofing.
- Infrastructure - Plays an increasingly critical role in providing information to vehicles regarding path planning and obstruction identification.

G. Meyer and S. Beiker (Eds.): AVS 2019, LNMOB, pp. 120–131, 2020.
https://doi.org/10.1007/978-3-030-52840-9_12

Each of these areas impacts the security of automated vehicles and should be considered to improve cyber resiliency.

2 Human Factors

Automated Vehicles will transport assets such as freight, goods, or humans. Therefore, the human factors shouldn't be overlooked. Let us describe two examples.

In scenario 1, a robot taxi transporting a user approaches a stop sign, but the vehicle does not stop and continues its course. If the Human Machine Interface (HMI) allows it, the user can take actions to inform the system (and the service provider) that a mistake happened[1]. The user could (i) notify that she felts that her safety was jeopardized or that the maneuver was uncomfortable, (ii) perform maneuver herself. In this example, the user can help improve the system to make it more robust. Of course, this opens up to poisoning attacks, where the users push bad notifications (similar to the creation of fake traffic jam on Google Maps). However, filtering techniques exist to help mitigate this risk.

In Scenario 2, GPS jamming is detected by the onboard security system[2], and the system decides to slow down in order to mitigate potential maneuver uncertainties. A passenger does not notice the attack and wonders why the vehicle reduces speed. Without proper HMI and considerations for human factors, this could result in lower trust in the system and the service overall. In this example, if the system is aware of the level of attention/knowledge of the user, it could (i) inform the user of the situation, (ii) ask for user's feedback.

Both scenarios demonstrate the need to study if and how users should be involved in case of a cybersecurity incident.

When such incidents happen, it is critical for drivers to respond properly and remain safe. Zhang et al. [1] performed a study to observe drivers' responses to their *non-automated non-connected* vehicle being hacked while driving in a simulated environment, and subsequently interviewed these drivers to gain deeper insights of their perceptions. Results showed that participants perceived and responded differently for each unexpected situation. Participants correctly identified the hack-induced issue and took according action when the hack caused a noticeable visual and auditory response. Participants preferred to be clearly informed about what happened and what to do through a combination of visual, tactile, and auditory warnings. The lack of knowledge of vehicle cybersecurity was obvious among participants. Hence, it is important to create a cybersecurity training and designed HMI that conveys relevant cybersecurity information for local and remote incident response.

An issue related to human factors is the evolution of HMI. Indeed, in an AV level 4–5, it is foreseeable that the vehicle will not have a system to manually maneuver the vehicle. This HMI evolution could affect what users can do to help the system mitigating incidents. Likely, the only two options will be to push an emergency button, and

[1] One could think of offloading current system engineer work (located in the backseat) to the user.

[2] And the system is considered to still be within its operational design domain.

a teleoperation of the vehicle. On the other hand, by removing maneuvering levers, we also diminish the risk of user interference. Indeed, if the user cannot affect the vehicle course/kinematic state, it can't counter unnatural –but safer— system behaviors. The design of cybersecurity-centric HMI for AV is an open challenge.

Tightly linked to the design of HMI, we identified a need to design standardized rules for human-AV interaction in case of cyber-incidents. The safety report published by ride-sharing companies could be extended to incorporate cybersecurity incident responses. First responders should be aware of the expected behaviors so that they can execute their missions safely and in timely manner. They will also have to be trained to face cyberattacks and protect evidence.

Teleoperation of AV is required for robot taxi service providers in order to resolve issues in real-time and minimize service disruption to the users. In [2], it has been shown that the communication latency (i.e. time it takes for a remote command to be received by the system under control) significantly affects operator's performance. Another aspect is what information should be sent to a teleoperator to mitigate the cyber-incident detected by the onboard security system, and how should this information be displayed in the security operations center.

As we can see, the cybersecurity responses are generated by machine (see next section) or humans, and hence, have to be designed with human factors in mind in order to be actionable and effective.

3 Machine

Proven to be effective in many areas such as object detection, image classification, and prediction, Machine Learning (ML) algorithms have found their way into many safety-critical applications. In AVs, for example, ML is used for object classification, path planning, and driver behavior, but because these algorithms were adopted so quickly, many developers did not fully consider the security implications that come with a new class of algorithms.

Generally speaking, ML models input sensor data and output a label. However, it has been demonstrated that ML models can be fooled by adversarial examples [3]. Adversarial examples are crafted by adding human-imperceptible[3] perturbations to inputs so that a neural-network-based classifier incorrectly labels them. Fast Gradient-Sign Method is a simple way to create adversarial examples in the example of perturbated traffic signs. The perturbation on each pixel is set to be of fixed distance, epsilon, hence guaranteeing that no pixel in the adversarial example is more than epsilon from the original image. Therefore, a human will not perceive the difference while the ML model will misclassify the object.

Researchers have uncovered flaws in many popular deep learning algorithms, causing them to misclassify objects, as shown in Fig. 1. Spoofing of object localization by creating patches has also been demonstrated. This process was improved upon by

[3] Note that some approaches for generation of adversarial example use unbounded distance, which can be perceptible by humans.

creating perception-invariant adversarial examples, created using full homography transformations of the adversarial example during training. The advantage of this approach is, when testing in the physical world, the adversarial example does not need to be perfectly parallel with the camera system. Instead, it can now have a certain degree of rotation in all three dimensions without compromising its effectiveness.

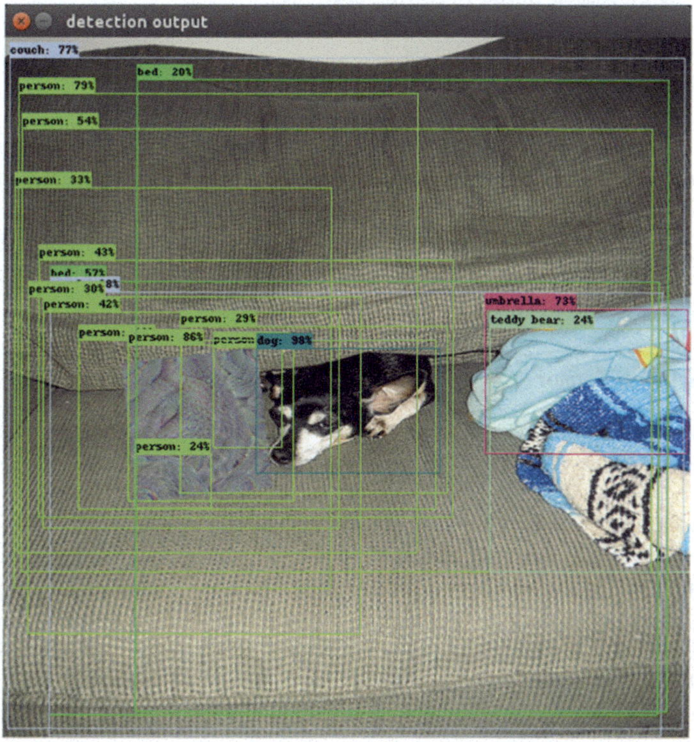

Fig. 1. Machine learning object misclassification

On top of being used for normal operation needs (e.g. object detection and classification), ML is used in security systems to detect network attacks. In this case, the ML model takes network packets in input and can detect anomalies. Therefore, with such potential weight into safety and security operations, it is not surprising that academic studies toward defeating ML-based systems are on the rise. The attacks aim at stealing, poisoning, or evading the ML models. Most of the current attacks are evasion attacks, where the attackers want to trigger false positive or false negative without affecting the ML model itself.

One of the main challenges of ML (especially deep learning) in AV is its lack of guarantee in safety, security, and predictability. Indeed, Mohseni et al. [4] highlighted the need to investigate the gaps ML models expose in existing engineering safety standards, such as interpretability and traceability into code, formal verification, and design specification [5]. To help with interpretability, one field of research is Explainable AI (XAI) [6], which advocates for production of explainable models while maintaining a high level of learning performance (prediction accuracy). To help with reliability, the Kayotee tool [7] was developed to systematically inject faults into software and hardware components to assess the safety and reliability of AVs to faults and errors. This tool can be extended to generate remote attacks on sensors as well, which will be discussed in the following section.

4 Sensors

Automated vehicles and Advanced Driver-Assistance Systems (ADAS) rely heavily on valid sensor readings in order to estimate their position and identify and track surrounding objects such as other vehicles. Attacks on sensors, whether remotely or through the vehicle's network, can create an unstable control system and require the driver to take over or risk a collision. This section presents examples of attacks on sensors, defenses, and open challenges to securing them.

4.1 Attacks on Sensor

Researchers have published and presented many instances of attacks on sensors. This section reviews some of these attacks and discusses the impacts.

4.1.1 Spoofing AV Sensors

Researchers have demonstrated spoofing of several popular sensors used on AV [8]. Spoofing attacks occur when a manipulated signal is sent to a sensor to imitate a valid signal. Spoofing can cause an object to appear that is not there, move an object closer or farther away than it actually is, or hide an object that is really present. Spoofing of LIDAR, Radar, Ultrasonic, and GNSS have all been published. GNSS spoofing is detailed in Sect. 5.2.1.

4.1.2 Camouflage

Sensor technologies can have significant benefits over human sensing. For instance, sensors have the ability to track many objects with high accuracy without being distracted; however, they also have inherent limitations. Radar, for example, has the benefit of seeing through rain, snow, and vehicles but struggles to track materials such as wood and fiberglass, which is why small boats that travel open seas should have a radar reflector on board. Ultrasonic sensors, which rely on high frequency acoustic signals, struggle to detect materials that absorb sound such as foam. Similarly, LIDAR can struggle to detect objects that absorb light such as items that are painted with a very deep black.

4.1.3 Sensor Obstruction and Blinding

Not all attacks on sensors require technical skill or are particularly difficult to perform. Covering LIDAR, camera, and ultrasonic sensors leaves them unable to sense their surroundings. Covering radar and GNSS receivers with conductive materials has a similar effect.

Lasers aimed directly at cameras are able to blind them, and all sensors can be saturated if a strong enough emitter is pointed at them. For example, a strong laser or dazzler pointed directly at a LIDAR sensor can prevent it from receiving legitimate responses. For radar, introduction of noise or a sine wave near the center frequency (e.g. 24 or 77 GHz) can also blind the system.

Obstruction attacks are incredibly easy to perform and can be difficult for drivers to notice. To avoid possibly disastrous outcomes from these attacks, it important for vehicles to check the sensors at startup and regularly during use. If a sensor receives no response when one is expected, the automobile should notify the operator (and hence, the importance of HMI as discussed in Sect. 2).

4.2 Defenses

While the published sensor attacks can have dire consequences, there are ways to detect or defend against some of them:

1. Filtering - Some LIDAR and radar sensors use sophisticated filtering to improve signal detection. These filters are a good defense against lower tech attacks and improve the quality of sensor output. For example, shining a low power laser at a LIDAR sensor from a distance may not show up in the sensor output due to signal filtering.
2. Sensor Monitoring – Automated vehicles can monitor sensor performance and track signal characteristics. Inconsistencies in sensor readings could indicate a problem with the sensor or indicate an attack.
3. Sensor Fusion – Compare data from multiple sensors. Unexpected discrepancies in data can indicate a problem with a sensor or an attack.
4. V2X Information Sharing– vehicles and infrastructure can share problems that are observed. As seen with some Tesla videos[4], there are rare instances that roadway setups could cause issues with sensors. Sharing incident information with others can help prevent similar issues in the future.

4.3 Open Challenges

Sensors are limited and all are susceptible to some level of blocking, blinding, camouflage, and spoofing. There is plenty of room for improvement in sensor monitoring, sensor fusion, and V2X information sharing.

5 Infrastructure

Infrastructure plays a significant role in modern vehicles as it provides them with connectivity (Road Side Units/Cellular/Wi-Fi). As vehicles move into AV level 4–5, communication with infrastructure will help keep the vehicle and passengers safe. Autonomous vehicles produce and process massive amounts of data. One LIDAR

[4] https://www.youtube.com/watch?v=fKyUqZDYwrU.

sensor alone can create millions of data points in one (1) second. Maintaining the integrity of data sent to the vehicle is required so that attackers cannot manipulate the vehicle. Automated Vehicles will rely on infrastructure and services, including communication networks (Cellular/Wi-Fi), Over-the-Air updates, for software, maps, and data servers, Global Navigation Satellite Systems (GNSS), and other physical features (e.g. ground characteristics used by localization systems). Manufacturers will likely manage fleets remotely and provide supporting services.

5.1 Examples of Infrastructure

5.1.1 Vehicle to Infrastructure (V2I)

Infrastructure and vehicles can communicate and provide each other with important information. Infrastructure can inform vehicles about travel times, lane and road closures, speed limits, traffic signs (e.g. stop or yield), traffic signals (e.g. red, yellow, green), and location of emergency vehicles. This information can be used to safely optimize travel routes, which benefits everyone. Each vehicle is also able to provide status information to infrastructure, such as speed of travel, location, acceleration, brake status, and collision detection. Pilot systems have demonstrated collision avoidance, combining vehicle information with data from infrastructure sensors to warn when pedestrians or vehicles are in the driving path but not necessarily visible to the driver.

5.1.2 Global Navigation Satellite Systems (GNSS)

GNSS systems provide localization estimates to receivers anywhere on the planet as long as they can receive a strong enough signal. These systems include GPS (USA), GALILEO (Europe), BEIDOU (China), and GLONASS (Russia). Public versions of these systems can provide location estimates with an accuracy of a few meters, which works great for route planning and position verification; however, they lack mechanisms to check integrity and are vulnerable to spoofing.

5.1.3 Over-The-Air (OTA) Software Updates

OTA software updates allow automobiles to securely update the firmware of ECUs after they are in customer's hands. Ten years ago, firmware updates on vehicles were unnecessary because exploitation of vulnerabilities required physical access. Similar to what we have seen with personal computers and cell phones, vulnerabilities will have to be patched as automobiles become connected or they will be susceptible to exploitation. Uptane (https://github.com/uptane) is an Attack-Resilient Open-Source Security Framework for Over-The-Air Software Updates. With the segregation of roles and multiple security keys, Uptane provides resilience against an attack even in the face of a security key compromise. Without compromising all of the security keys (online and offline), attackers are prevented from injecting their malicious code into a remote software update. What was once an easy target for remote exploitation is instead a well-protected secure mechanism for OEMs and suppliers to provide the best functionality to their customers quickly and without extensive down-time (Fig. 2).

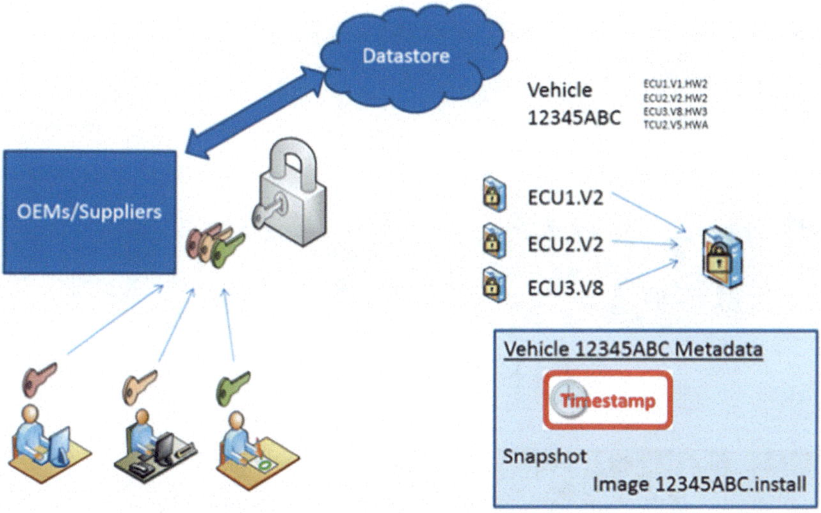

Fig. 2. UPTANE overview

5.2 Cybersecurity Attacks on Infrastructure

This section reviews two examples of attack associated with infrastructure and considers the potential effects.

5.2.1 GNSS Spoofing

Public versions of GNSS lack integrity verification mechanisms, leaving the signal vulnerable to spoofing. Several publications have documented exploitation of this vulnerability, and the reduced cost of software defined radios has lowered the barrier to entry. A $400 software defined radio and a laptop are the primary hardware components needed to spoof GNSS. It is common for GNSS to be blocked, distorted, or reflected by buildings, tunnels, overpasses, or other structures.

GNSS is commonly used by automobiles for route planning and for localization estimates on connected and autonomous vehicles. Researchers created a system that allowed to manipulate the GNSS signal remotely in real time on standard and automated vehicles. The GNSS used by a vehicle's navigation system was manipulated to have the vehicle believe it was in another country while driving in the US.

GNSS spoofing effects on an AV were also demonstrated. The vehicle was outfitted with a SwRI autonomy package and configured to drive by GNSS waypoint, which maximized the reaction to manipulation. It is common to blend GNSS with other localization estimates, which could reduce the effects. The different manipulations and their effects are detailed below.

The first manipulation was to insert an offset. This step provided the ability to force lane changes, drive the car off the road, or cause it to turn early or late. In Fig. 4, the modified GNSS forced the vehicle off the road. Figure 3 shows the vehicle driving with unmodified GNSS.

Fig. 3. Offset applied mid-route

Fig. 4. Nominal run (no attack)

Fig. 5. Halt attack demonstration

Researchers also showed the ability to stop the GNSS signal without losing a fix, which causes the controls system to become unstable due to lack of accurate positional feedback. Figure 5 shows the altered GNSS and the unstable vehicle response while Fig. 6 shows the unmodified GNSS.

Fig. 6. Nominal run (no attack)

Note that if GNSS were lost, the vehicle would stop because it recognizes an issue. With these attacks, the vehicle always has a solid GNSS signal, and the system infers it is able to continue driving.

5.2.2 (Potential) Manipulation of Connected Vehicle Data

Several connected vehicle pilot programs have been deployed in small numbers. Connected vehicle technology has the potential to dramatically reduce accidents by providing drivers with collision warnings. When these programs are deployed in mass, they will collect data from a large number of vehicles. Without a rigorous verification process, packets could be spoofed and the systems tricked into thinking major traffic incidents have occurred, triggering warnings to drivers.

5.3 Open Challenges

Open challenges for infrastructure include the following:

- Authentication of data from vehicles is needed so that data can be trusted and used to make or guide decisions involving safety. Overcoming this challenge will require a careful balance of privacy versus data authenticity.
- SOTA – not deployed in mass. Systems cannot remain secure for their expected life without ability to patch vulnerabilities.

- Camouflage – must incorporate sensor fusion for verification/validation of both localization estimation and object detection.

6 Conclusion: All Together Against Cyberattacks

Figure 7 shows the interaction between Sensor, Machine, Human and Infrastructure in the AV ecosystem. Sensors are reading their environment and send that data to the Machine (1). The Machine processes the data in order to maneuver and inform the Human (2). The Human (by her level of attention, user's profile, or reactions) can affect the sensor and the machine (3). The infrastructure (a, b) can provide input to Sensor and Machine through over-the-air update, geolocalization services for example. It is evident that a single weak link has the potential to disrupt the cycle.

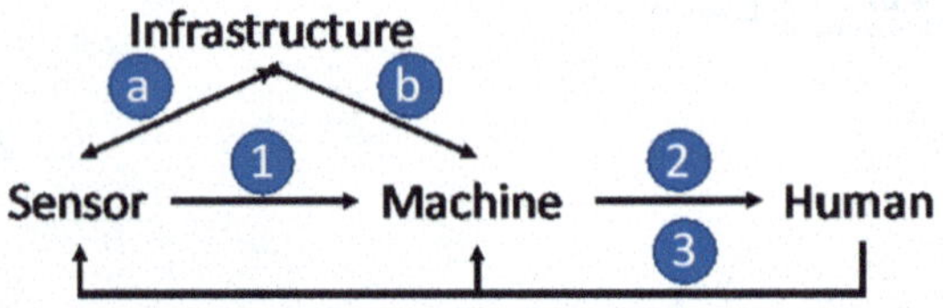

Fig. 7. Interplay between AV's ecosystem components

As discussed in the earlier sections, each component has its set of attack surfaces, system requirements, and open challenges. One could notice that, when taken individually, none of the components consider the other ones in order to identify, protect, detect, respond and recover from cyberattacks. To some extent, this behavior makes sense. For example, a sensor supplier does not have control over the infrastructure or the machine it is sending data to, and hence, can't directly influence their respective protection profile. Nevertheless, we believe that stakeholders cooperation is paramount to build resilient AVs. As the certification framework around automotive cybersecurity is shaping, we can envision that interoperability (e.g. in terms of reporting, evidence collection) and coordination (e.g. for identification, reaction and response) will improve. Detecting an attack at a component-level is good but spreading the knowledge within the ecosystem is better. Indeed, the other components can analyze the attack and decide for themselves how to adjust parameters under their control. To conclude, even if standardization and regulation of AV cybersecurity could significantly improve the overall system resilience, we encourage stakeholders and academics to not wait and continue their effort in designing an end-to-end security solution instead of a compilation of point solutions. Finally, because of its safety/efficiency/privacy impact, cybersecurity should be by design and not an afterthought.

Acknowledgement. We would like to acknowledge John Moore (Ford Motor Company) for Reviewing Paper, and Harold (Abe) Garza and David Chambers (Southwest Research Institute) for their contributions to the adversarial learning section.

References

1. Zhang, F., Petit, J., Roberts, S.C.: A simulator study on drivers' response and perception towards vehicle cyberattacks. In: Proceedings of the Human Factors and Ergonomics Society Annual Meeting, vol. 63, pp. 1498–1502 (2019)
2. Neumeier, S., Wintersberger, P., Frison, A.K., Becher, A., Facchi, C., Riener, A.: Teleoperation: the holy grail to solve problems of automated driving? Sure, but latency matters. In: Proceedings of the 11th International Conference on Automotive User Interfaces and Interactive Vehicular Applications, pp. 186–197 (2019)
3. Sitawarin, C., Bhagoji, A.N., Mosenia, A., Chiang, M., Mittal, P.: DARTS: deceiving autonomous cars with toxic signs. arXiv preprint arXiv:1802.06430 (2018)
4. Mohseni, S., Pitale, M., Singh, V., Wang, Z.: Practical solutions for machine learning safety in autonomous vehicles. arXiv preprint arXiv:1912.09630 (2019)
5. Salay, R., Queiroz, R., Czarnecki, K.: An analysis of ISO 26262: using machine learning safely in automotive software. arXiv preprint arXiv:1709.02435 (2017)
6. Gunning, D., Aha, D.W.: DARPA's explainable artificial intelligence program. AI Mag. **40** (2), 44–58 (2019)
7. Jha, S., Tsai, T., Hari, S., Sullivan, M., Kalbarczyk, Z., Keckler, S.W., Iyer, R.K.: Kayotee: A fault injection-based system to assess the safety and reliability of autonomous vehicles to faults and errors. arXiv preprint arXiv:1907.01024 (2019)
8. Petit, J., Stottelaar, B., Feiri, M., Kargl, F.: Remote attacks on automated vehicle sensors: experiments on camera and LiDAR. BlackHat Europe (2015)

Part V Policy and Planning

Preparing for Automated Vehicles and Shared Mobility: The Existential Questions

Katherine Kortum(✉)

Transportation Research Board, 500 5th Street NW,
Washington, DC 20001, USA
kkortum@nas.edu

Abstract. This session discussed mostly long-term questions around automation and shared mobility, summarizing the work that the TRB Forum on Preparing for Automated Vehicles and Shared Mobility has done in its year and half of existence. Panelists summarized workshops that the Forum members held during 2019 and described the key workshop takeaways to the audience, and then engaged the audience in a discussion about what research topics are still needed.

Keywords: CAV · Automation · Transportation planning · Research · Economic forecasting

1 Introduction

The 2019 Automated Vehicle Symposium, organized by the Association for Unmanned Vehicle Systems International and the Transportation Research Board (TRB), included a Breakout Session titled *Preparing for Automated Vehicles and Shared Mobility: The Existential Questions.* This session comprised a panel of experts discussion the results of workshops held by the TRB Forum on Preparing for Automated Vehicles and Shared Mobility over the course of 2019.

In 2018, TRB launched this Forum to bring together public, private, and research organizational partners to share perspectives on the critical issues surrounding the deployment of automated vehicles and shared mobility. The Forum's key emphasis is on discussing, identifying, and facilitating fact-based research that is needed in order to deploy these technologies in a manner and timeframe that informs policy to best meet long-term goals. During 2019, the Forum held multiple workshops on several "existential questions," allowing for deep discussions of the topics and identification of research still needed on each topic. The Automated Vehicle Symposium session, along with this chapter, summarize these discussions and highlight the key research needs from each 2019 event. For each workshop, one of its organizers provided the summary.

G. Meyer and S. Beiker (Eds.): AVS 2019, LNMOB, pp. 135–143, 2020.
https://doi.org/10.1007/978-3-030-52840-9_13

2 Transitioning Toward Shared and Automated Vehicles

Two megatrends are currently shaping transportation throughout the world:

- Transportation issues requiring a change from business as usual: rapid urbanization, safety, demographics, emissions, access, congestion, efficiency
- Digital transformation trends: e-commerce digitization, platforms, and connected devices, shared mobility scaling, mobility as a service/mobility on demand.

These trends are upending traditional notions of transportation and bringing unprecedented opportunities and challenges to travel modes, urban form and land use, propulsion, employment, safety, data management, and ultimately governance. This shift will not happen all at once, and it is clear it will be a transition–one that will have visibly different geo-spatial and temporal effects depending on the physical, political, and social situation of each urban region.

This TRB Forum on Preparing for Automated Vehicles and Shared Mobility focuses on sharing specifically in the context of AVs, which includes passenger, delivery/commercial and municipal services, shared fleets, and shared rides (sequential and concurrent sharing), along with ground, sea, and air systems.

Data technology companies are dominating market share of traditional industries with a platform approach. By linking producers of services with customers who purchase those services, the data transaction exchange creates value for the company, producer, and customer; it can scale to all parts of the economy.

The key question raised in these discussions was how to bring governance into the era of digitization, so they are able to manage demand and supply digitally. At the same time, we must be correcting for the existing inequities in our system, working closer with public and private sector partners, and developing the infrastructure needed to ensure that the transition to SAVs minimizes the potential challenges and takes advantage of those opportunities that may come out of this transportation transformation.

This transition and its effects will not happen tomorrow, and there are several unknowns. The set of discussions in this section is intended to map out what we know, what we do not know, and what we need and want to see happen to get to the point where SAVs are on the roads and providing overall benefits to society.

Prachi Vakharia of Steer reviewed a workshop about the transition period to automated vehicles. The workshop identified what research is needed to transition to the vision of full automation, and a few key research questions identified for regulation, equity, and land use/streets are below.

Regulation research

- Responsibility for regulating the driving task: clarifying federal vs. state roles
- Safety tipping points regarding levels of penetrations of connected and automated vehicles (CAVs)
- Standard methodologies for state/local agencies to act on results from demos and other data/information

Equity research

- What are transportation cost and travel time implications for different groups?
- Public engagement and a market research/gap analysis. Most transportation users are not currently targets of private sector market research
- How do we ensure equitable access to digital platforms?

Land use/streets research

- Design for a transitional period which will operate in a mixed environment and require consumer education
- What are the best uses of existing rights-of-way? Dedicated lanes, managed curb space, EV charging networks, etc?
- Pickup/dropoff (PUDO) zones - waiting time and fees, among others. This can build upon existing design practices, using airports as an example, and can build upon National Association of City Transportation Officials (NACTO)/Institute of Transportation Engineers (ITE) work.

3 The Importance and Role of Connectivity

Jeff Lindley of the Institute of Transportation Engineers described the importance of connectivity, even in a world so focused on automated vehicles. Highlights of connectivity discussions are summarized below.

There is considerable popular press about autonomy, but only industry insiders are talking about connectivity. More needs to be done to create consumer awareness of the differences between and the benefits of both connected and automated vehicles. Although automated vehicles and connected vehicles are often grouped together, their situations are different in terms of technological maturity and uncertainty.

It is important to ensure that both the deployment of both connected and automated vehicles, in addition to shared mobility services, focus on a user perspective, not on a vehicle or infrastructure perspective.

Connected automation is critical to enable the transportation system to function most safely and effectively. Two critically important reasons for connectivity include redundancy and the ability to see what sensors cannot. On-board sensors work for immediate vicinity sensing to support collision mitigation and avoidance; cellular communications and dedicated short range communications (DSRC) can support a broader range of applications and over longer distances (such as avoiding collisions due to overtaking maneuvers on high-speed rural two-lane roadways). Also, pricing of streets and managing of curbs will require connectivity.

Connectivity may require a different legislative and regulatory framework than exists today, particularly with respect to public and private sector relationships and to funding and risk assignment and liability.

Consistency and interoperability are critical to the successful deployment of connectivity. Standards development processes may need to be streamlined and accelerated.

Vehicle manufacturers are concerned over state DOT inconsistencies in deploying DSRC. On the other hand, state DOTs cannot be dependent on proprietary clouds of data housed within each vehicle manufacturer.

There is significant fear regarding the liability of a failure in connectivity. Failures might include a complete breakdown in the connectivity or the speed being too slow, resulting in injury or death. Even if the technology works perfectly, approximately one-third of all crashes involve driver impairment, and impaired drivers cannot be expected to be able to react to the alerts provided by a vehicle.

The technology debate will rage on, but it is important to focus on the use cases. Each use case is likely to require something different.

The number of devices on the road should not be the sole measure of readiness. Policy, maintenance, and workforce issues matter as well, and pilot projects are the means to determine how these will work. In addition, connectivity is not free. Industry and the public sector need a better understanding of overall costs of the technology, including maintenance costs, before the system is truly ready to accept widespread connectivity.

There will continue to be more questions than answers for the foreseeable future, and ongoing research will be needed.

Research questions around connectivity include:

- What are the incremental benefits of deploying connected AVs beyond deploying just (unconnected) AVs?
- How do connectivity needs and benefits change between very near, mid-distance, and long distance applications?
- How can the needs of vulnerable users (e.g. pedestrians and bicyclists) be accommodated without requiring them to bear the full responsibility of always being "connected"?
- What viable options exist for managing and providing access to the data that will be generated in a fully connected environment without compromising privacy or creating cybersecurity risks?
- What is the consequence of sharing information such as a wrong-way driver with human drivers or human AV safety operators?
- How do humans react when they receive messages that are time critical and potentially life-threatening, and how does this vary depending on how the message is provided?

4 Guiding Principles for Connected Infrastructure Supporting Cooperative Automated Transportation

Collin Castle of Michigan DOT represented the CAT Coalition and discussed its latest efforts for guiding principles. Infrastructure Owners and Operators (IOOs) need these Guiding Principles (GPs) for connected infrastructure supporting CAT to facilitate collaboration, educate the workforce, support interoperable deployments, and inform the public. In the near term, the GPs reflect the consensus direction of the IOOs and can support impact assessment of CAV developments. Over the longer term, GPs are intended to give IOOs maximum institutional flexibility while working together to develop and deploy CAT strategy, standards, data exchange, best practices, and public information, among others. The principles are below.

1. Automation. Support increased vehicle automation to improve traveler safety, mobility, equity, and efficiency.
2. Data. Achieve a connected vehicle ecosystem that enables reliable, secure V2I data exchanges in order to support cooperative automated transportation.
3. Telecommunications. Protect and utilize the 5.9 Gigahertz (GHz) spectrum designated for "operations related to the improvement of traffic flow, traffic safety and other intelligent transportation service applications" (FCC).
4. Multi-Modal Operations. Develop CAT strategies that enhance existing transportation system operational capabilities.
5. Multi-Modal Collaborations. Collaborate and communicate with CAT stakeholders in the planning, testing, and demonstrations of CAT applications to support eventual interoperability and to achieve positive impacts on safety, mobility, equity, and efficiency.

5 Potential Impacts on the Roles of Different Levels of Government and the Private Sector

Virginia Reeder of the I-95 Corridor Coalition described a workshop that finished only hours earlier, focused on the potentially changing roles of government and the private sector. The workshop asked participants to role-play to consider what other sectors needed from the different players in the AV world.

Key takeaways from the workshop included the following:

Current roles will pave the way, but partnerships and relationships will need to change. It is inevitable that the new technology associated with AVs and shared mobility will impact our transportation network, options, behaviors, and needs. While many traditional roles may not shift drastically (e.g. infrastructure operator, transit service provider, etc.), each party will need to listen and collaborate with a new set of partners to ensure that we realize the benefits, work through the challenges, and solve the inevitable issues.

It is key to acknowledge what we don't know. There has been a great deal of hype about the advent of automated vehicles and shared mobility. Pieces of the envisioned future already exist, but large questions remain about the technology, implementation, policies, and the many expected outcomes and impacts. It is incumbent on all stakeholders to openly acknowledge that the path forward is not clear, and the actions of different partners can affect the way things unfold. This shared understanding will allow for the most comprehensive and creative solutions.

Everyone wants guidelines, frameworks, etc. With all of these unknowns, each entity is looking for guidelines and frameworks that establish key roles and responsibilities as well as consistent yet flexible regulatory approaches within which to move forward with technology development, policy adoption, and new responsibilities. Understandably, those who have traditionally provided this structure are hesitant to adopt practices in the face of an unknown future and with the concern that too much regulation could hinder innovation. In addition, most states have indicated a preference for an open format that still protects data privacy, and the structure of such a system is not yet clear.

Consistency and standardization are critical, but it is unclear who should establish these. There may need to be a few new seats at the table with a slightly different set of perspectives in order to find that balance between a set path and room for playing in the sand box.

Vehicle manufacturers will continue to push agencies to improve existing traffic control devices. Where road infrastructure is inadequate or not reliably maintained, automated driving system (ADS) operated vehicle manufacturers may limit the operational design domain (ODD) of their vehicles and/or geofence difficult roads/intersections in order to avoid them.

6 Economic Implications of Automated Vehicles and Shared Mobility

King Gee of AASHTO led a workshop on economic implications of AVs. The workshop asked participants to consider the research we still need on the economic implications of AVs that would inform actions to maximize benefits and mitigate negative impacts. At the systems level, players need to test and validate key assumptions that are commonly in macroeconomic studies and better understand how technical feasibility and business models are applied in specific scenarios under different environments (politics, demographics, density, etc.). There are also questions about what economic and safety benefits are available at different levels of market penetration. It is also important to develop a shared vocabulary and common frames-of-reference for concepts such as Mobility-as-a-Service and ancillary services.

Economic impacts of transformational changes are difficult to forecast. Economic forecasts of previous transformational changes have missed outcomes, often by a wide margin, and these impacts have usually been underestimated. As a result, decisions have been made on a "common sense" basis, with relatively little reliance on economic analysis. Further, real economic impacts have usually been non-linear.

Clear funding and business cases will be important for the success of AVs and shared mobility. There is currently a great deal of variation around these cases and it is not clear where companies and the public sector will land.

Research is still needed in the following areas:

- Efficacy of key assumptions
- Modeling for specific scenarios
- Impacts of user acceptance

- Economic and safety benefits that accrue at different levels of market penetration?
- Economic impacts when good options are provided to the transportation disadvantaged
- Relationships between economic impacts and impacts on other policy issues
  - Environment, congestion, sprawl, equity, sustainability
  - Jobs
  - Transit
- Impacts of pricing tools
  - VMT, parking, curb space access
  - Transaction cost factors
  - One payment systems
  - Mobility-as-a-Service (Maas)

7 Potential Impacts on Our Traditional Research Processes and Programs

Finally, Mark Norman of the Transportation Research Board discussed the speed of the research process and how researchers have struggled to keep up with the speed of development in transportation. The need for speed depends on the situation and risk level; some situations still require longer research efforts but others that are lower risk can be accomplished more quickly.

Discussion on the topic included the following insights:

- Automated vehicles, shared mobility, and other transformational technologies in transportation provide a unique opportunity to make significant advances in meeting societal goals
- Success is far from assured – we have more questions than answers. Research is the key.
- Time is short, as these technologies and deployments are advancing, and often changing, rapidly
- The "need for speed" in the associated research depends on the issue and/or situation
- Answering a series of questions can help researchers to determine a targeted time frame appropriate for the specific research in question (see list below)
- Options do exist to enable our research projects and processes to provide needed answers in a timely manner while still protecting the credibility of our research (see list below)
- A number of these options have been employed by others, but more models are needed
- Collaboration among the public, private, and academic sectors is key to meeting the twin objectives of providing needed answers in a timely manner while still protecting the credibility of our research

- There are a number of steps that TRB can take, including providing more opportunities for collaboration among the stakeholders through TRB convening activities such as this Forum, conferences, standing committees, and research panels

Mark described a variety of speed-up steps to consider at each stage of the research process. Highlighted steps include:

- Reducing any administrative burdens along the way, including having a standing pool of researchers and reviewers ready to go
- Breaking research projects into smaller pieces that can be completed and disseminated more quickly
- Preparing Requests for Proposals that are focused on the outcome but do not specify the process for researchers to take
- Awarding research projects with shorter phases and instituting penalties for not meeting deadlines
- Work more closely with communications and public affairs staff to package the findings for different audiences

8 Suggested Action Items

Panelists and participants discussed a wide variety of research needs, only some of which are listed here. For more information, see the e-circulars, one summarizing each workshop, which TRB will publish in coming months. The first of these e-circulars are available as described below:

- *TRB E-Circular 247: Mini-Workshop on the Importance and Role of Connectivity* is available at http://www.trb.org/Main/Blurbs/179187.aspx.
- *TRB E-Circular 252: Mini-Workshop on the Transition Toward Shared Automated Vehicles* is available at http://www.trb.org/Publications/Blurbs/179691.aspx.
- *TRB E-Circular 253: Impacts on Traditional Research Processes and Programs* is available at http://www.trb.org/Publications/Blurbs/179824.aspx.

As of November 2019, the remaining two summaries are in production.

Participants suggested that continued collaboration is key, and that the Automated Vehicle Symposium and other TRB meetings are one potential resource for such collaboration. More information about past and future Automated Vehicle Symposia can be found at https://www.automatedvehiclessymposium.org/home. More information about the TRB Annual Meeting can be found at http://www.trb.org/AnnualMeeting/AnnualMeeting.aspx, and other TRB-related events can be found at http://www.trb.org/Calendar/Calendar.aspx.

Continued testing and demonstrations are important. Eventually, panelists and participants encourage those engaging in such testing to develop syntheses of the findings and recommendations from each test so that the lessons can be applied more broadly.

Acknowledgements. The author wishes to thank the many individuals who helped to organize these sessions throughout 2019. These include Susan Shaheen, University of California at Berkeley; Tim Papandreou, Emerging Transport Advisors; Jeff Lindley, Institute of Transportation Engineers; Ed Straub, SAE International; Mark Norman and Adrienne Archer, Transportation Research Board; Abbas Mohaddes, Econolite; Patricia Hendren and Virginia Reeder, I-95 Corridor Coalition; Scott Schmidt, Alliance of Automobile Manufacturers; King Gee, American Association of State Highway and Transportation Officials; and Art Guzzetti, American Public Transportation Association.

Benchmarking Automated and Autonomous Vehicle Policies in the United States

William Riggs(✉)

University of San Francisco, 2130 Fulton Street, San Francisco, CA 94117, USA
wriggs@usfca.edu

Abstract. Autonomous vehicles (AV) have dramatic potential not only to reshape cities. This research surveys 602 US cities to investigate how they are preparing for urban autonomy. To conduct this evaluation, roughly 20 key indicators are evaluated. Based on the evaluation most cities in the United States do not have AV policies. Of these the majority have some sort of white paper or policy acknowledgement. Key themes are in management (of transit, systems, parking, curb, data, etc.) and design (primarily streets and electric vehicle infrastructure), but not on travel behavior or demand management. This offers an opportunity for planners, engineers, policy makers and innovators in the coming years.

Keywords: Urban technology · Autonomous vehicles · Autonomy · Transportation · Benchmarking

1 Introduction

Over the past year there have been many technological revelations that have made it clear that autonomous technology will become increasingly prevalent in our communities. Despite optimism about the technology [1–4], there is a high degree of uncertainty about how it will outlay in urban environs, and there is very little work to planning and decision-making–especially given that land use and transportation actions have long inertial properties [5, 6].

Academics have suggested that policy action can be taken to support and adapt more quickly to disruptive innovations [1–4, 7], yet research has indicated that only 2 of the 25 largest metropolitan areas mention autonomous or connected vehicles their planning documents [8]. More recent reports from Bloomberg list approximately 30 cities globally with distinct plans; 17 of which are in the US.

This research extends that work, surveying 602 US cities and investigating how they are preparing for automation. The study benchmarks local planning policy response to autonomous vehicle technology in the United States where the pace of technology appears to be exceeding the pace of empirical planning. The work was completed as a part of the 2018 Web Benchmarking Study [9] which has been conducted biannually since 2014, and underpinned with expert surveys at the 2019 TRB/AUVSI Autonomous Vehicle Symposium and 2019 Autonomous Vehicles and the City conference hosted by the University of San Francisco.

G. Meyer and S. Beiker (Eds.): AVS 2019, LNMOB, pp. 144–160, 2020.
https://doi.org/10.1007/978-3-030-52840-9_14

The benchmarking methodology was based on a process to survey technological change in by cities and governments originally developed by Riggs, Steins and Chavan [9–12]. It was extended to account new emerging and disruptive technology, particularly in the areas of transportation and housing technology, namely autonomous vehicle and disruptive transportation policy.

This chapter documents how cities are responding to changes new and emerging mobility—including both transportation network and autonomous vehicle policy. It assesses the information produced from a large data set and then develops cases. In doing so, best practice policy is identified, including curbside policy, roadway thinning and parking removal. The ultimate goal is to offer planners, engineers and policy makers a framework to appropriately respond to the uncertainty of autonomous vehicles: providing a roadmap for local planning policy in such a time of uncertainty.

2 Background

New and disruptive transportation technologies not only influence the way people move throughout their communities and interact with one another, but many planners predict they will influence the way planners and policy makers interact with their citizens—much with increased reliance on e-government [13, 14]. Internet-enabled mobile devices incorporating GPS have allowed for location-based that can increase awareness of user activity, movements, and behaviors in real-time conditions and specific contexts [15].

These real-time conditions create an opportunity for a more legible urban landscape for the citizen, thus creating more efficient and sustainable mobility patterns throughout an urban environment [16]. At the same time, benchmarks of the use of participatory technology in cities across the United States, evaluating recent changes in how the top 500 cities in the US, suggesting high mobile use but very few mobile platforms [12]. In this new (primarily mobile) technology world, governments must plan, communicate and engage with citizens [14, 17, 18]. As shown in Fig. 1, tools that provide *interactive* and *transactive* platforms can help spur community innovation while democratizing the policy-making process [17, 19]. Specifically, as suggested by Riggs and Gordon *technology is becoming more transactive and cities must become more adaptive.*

This is particularly acute in the area of new and emerging forms of transportation—especially focused on autonomous vehicles [8, 6]. AVs present new opportunities to connect individuals to jobs and change the way cities organize space and optimize trips [2, 20]. While much of this is predicated on how much car manufactures are able to transition, or perhaps wean, customers from the idea of owning a car (a shared/subscription-based assumption that embedded throughout this text), many car companies are working on this and to challenge the centrality of the automobile in our public realm [21]. Cities are already seeing increases in driving caused by transportation network companies like Uber and Lyft, and changes to how many travel because of mobile phones and e-commerce [22, 23].

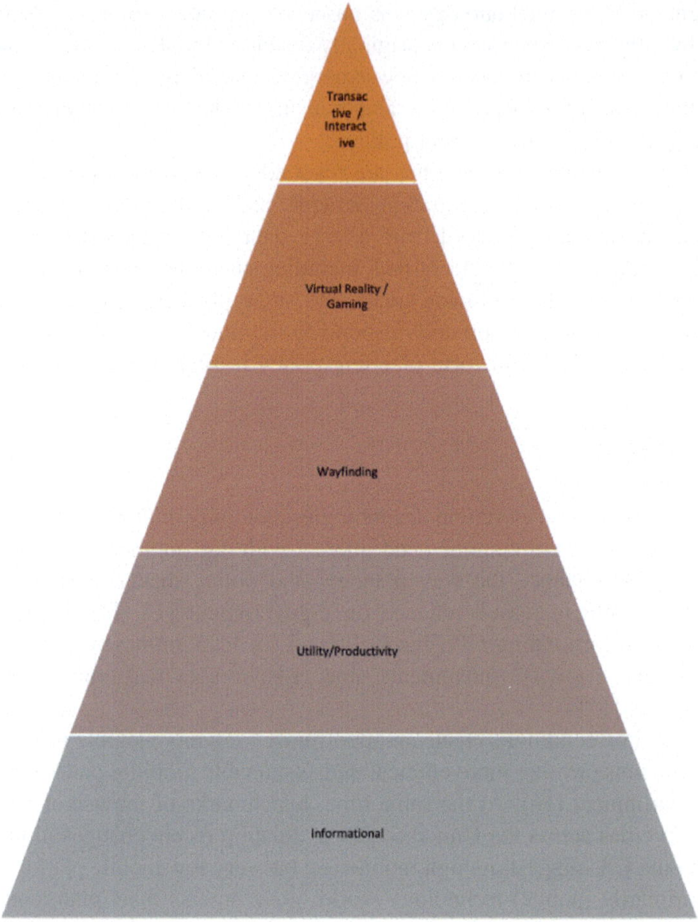

Fig. 1. A hierarchical taxonomy of urban technology. (Adapted from Riggs & Gordon [12, 17].)

Because on-demand transportation services allow for convenient point-to-point mobility, some say they have the potential to reduce automobile ownership in urban areas with the potential to facilitate first and last mile connections [24, 25] and compliment transit [26]. Other work has suggested that this smart and connected form of mobility actually reduces public transit ridership and increase VMT [22], and created complicated pick up and drop off issues [27].

The TNC revolution is compounded and is underscored by moves toward autonomous vehicles [21]. As self-driving cars are getting smarter and merging with our devices, there are clear opportunities to shape advances in transportation, to and harness them to reshape cities and improve the socio-economic health of cities [28]. There are opportunities to reduce collisions and improve access to healthcare and to connect individuals to jobs and change the way cities organize space and optimize trips [2, 8].

That said, there are also challenges and little understanding of how travel behavior and streets will change in this brave new world. Some work has used chauffer-based experiments to try to test behavior [29] –providing the best simulation of this technology thus far–and many other high-level policy documents have made suggestion and prediction that range in type and scale—both time and geographic (Anderson et al. 2014; Isaacs 2016; Litman 2014; Airbib and Seba 2017). Work by Appleyard, Riggs, Schlossberg, and others has explored the urban design and built environment possibilities presented by technology [30–33]. Specifically, much of this suggests that streets can be rethought with more space allocated to non-automotive modes. The bulk of this work predicts that this disruptive and accelerated transportation environment will continue and even accelerate, yet at the same time technological change is outpacing local policy [7].

This lack of focus on policy and decision-making in this high uncertainty environment is unique–especially given that land use and transportation actions have long inertial properties [5, 6]. In terms of urban policy, some of the best work shows that approximately 8% of the largest metropolitan areas mention autonomous or connected vehicles their planning documents [8]. Recently NACTO and the APA release broad policy guidance [34, 35], and organizations like Bloomberg, McKinsey and others have kept pace of market changes [36, 37]. Yet little is known about the way individual cities are responding to AVs and what specific actions they are taking.

3 Methodology

To conduct this evaluation, roughly 20 key indicators were evaluated for 602 cities across the United States between April and May of 2018. This methodology for benchmarking websites was based methods used to establish conducted on the technological change in by cities and governments by Riggs, Steins and Chavan [10, 11]. It was extended to account new emerging and disruptive technology, particularly in the areas of transportation and housing technology, namely autonomous vehicle and disruptive transportation policy.

Table 1. Relevant variables and assessment criteria for urban technology-related attributes gathered as a part of the benchmarking study.

Variable	Attributes and Measurement Strategy
Open Data	• Does the city have an open data portal or is there evidence of open planning data? • Record: Yes/No (or Unable to ascertain); If Yes, Document Link; If No, means we were simply unable to find
Traffic Routing Policy	• Does the city have policy or documentation on their website on how to handle traffic routing through neighborhoods? • Google search or city search engine on terms (Waze, Google Maps, etc.); Record: Yes/No (or Unable to ascertain); If Yes, Document Link; If No, means we were simply unable to find

(*continued*)

Table 1. (*continued*)

Variable	Attributes and Measurement Strategy
Transportation Network Company Policy	• Does the city have policy on their website on Transportation Network Companies (TNCs) like Uber and Lyft? • Google search or city search engine on terms Uber, Lyft, TNC; Record: Yes/No (or Unable to ascertain); If Yes, Document Link; If No, means we were simply unable to find
Autonomous Vehicle Policy	• Does the city have an autonomous vehicle (AV) policy on their website? • Google search or city search engine on terms; Record: No/Yes, mention (in policy)/Yes, white paper/Yes, Ordinance/Yes, General Plan; If Yes, Document Link; If No, means we were simply unable to find
CTO/CIO	• Does the city have a Chief Technology or Information Officer (CTO), or Chief Innovation Officer? • Make sure you make a distinction between the information/tech and innovation. Search the cities website. This will likely be in an org chart • Potential Answers: No; Yes, only CTO; Yes, only CIO; Yes, both

In terms of data details, web data was gathered using the BuiltWith tool. Other data for new mobility and smart cities was classified by trained data gatherers. They assessed if planning websites have policy, plans or language in 5 key areas: Open Data; Traffic Routing; Transportation Network Companies (TNCs); Autonomous Vehicles; Airbnb/Vacation/Short-term Rentals. We also evaluate if cities have a Chief Information/Innovation Officer (CIO) or Chief Technology Officer (CTO). For these "smart city" factors the assessment rules are provided in the Table 1. Policy analysis policy was classified using thematic coding and validated based on subsequent questions asked to subject area experts at two conferences: AVS and the Autonomous Vehicles Symposium.

For the cities we evaluated our longitudinal unit of analysis (the top ~500 cities in the US based on 2012 Census ACS 1-year data) remained the same, although we continue to add a random sample of cities below the 60,000-person ACS threshold, using Census 2012 3-year estimates. There were a total of 2,143 municipalities with estimates in the 2012 3-year data. A histogram of the cities is provided in Fig. 2.

Typically, the most current data is found in the 1-year estimates, although this data lacks the precision of 5-year estimates. Using the 1-year figures was appropriate since the goal was to benchmark the up-to-date characteristics of web technology in planning organizations throughout the US. The 1-year ACS data also offered the capability of reputable and repeatable sourcing method that was tied to existing demographic data in the communities evaluated.

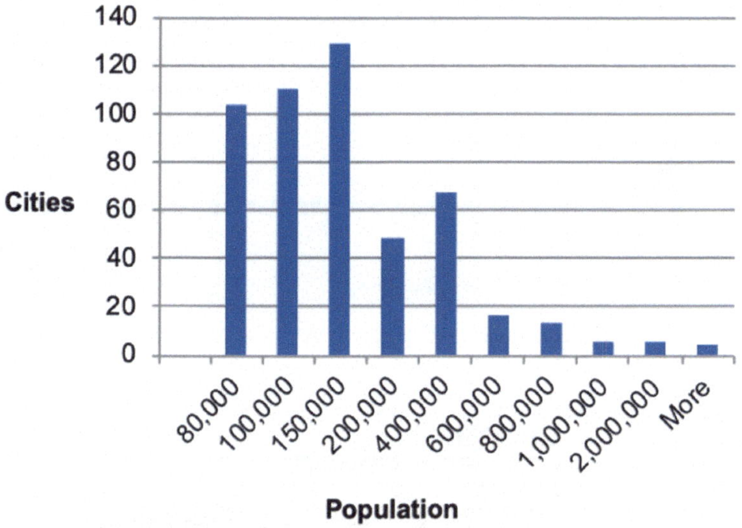

Fig. 2. A histogram of the population of the cities assessed.

For the purpose of this assessment, given the distribution, these are classified into cohorts for analysis. The cohorts are based on size grouping modeled after the *Research Brief on America's Cities* [38]. Cities less than 100,000 are classified as small. Those with populations between 100–199,000 are medium, and those with populations greater than 200,000 are large. This is also consistent with other work that shows larger cities have different socio-economic dynamics than cities that are smaller – a factor that could contribute to differences in technology adoption and use [39–41]. The sample (N = 602) was statistically significant with at the 95th confidence interval and a margin of error of ±3.93%.

4 Results

Based on the evaluation 459 of the top 602 cities in the United States do not have AV policies of any kind. Seventy-five (75) out of 602 cities sampled have AV policy (or about 12%). Of these 29 have an ordinance or general plan (approximately 5%). Seven (7) have a white paper and 34 have some kind of mention in their plans. This breakdown is illustrated in Fig. 3. As can be noted in Figs. 4, 5 and 6 there is very little correlation with cities having AV policy and engaging in other technology best practices. Cities that are tackling AV policy may not have things like CTOs or OpenData portals.

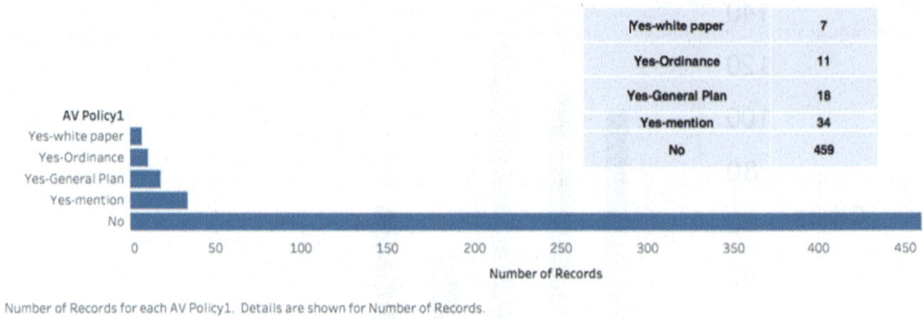

Fig. 3. Number of cities with AV policy by type.

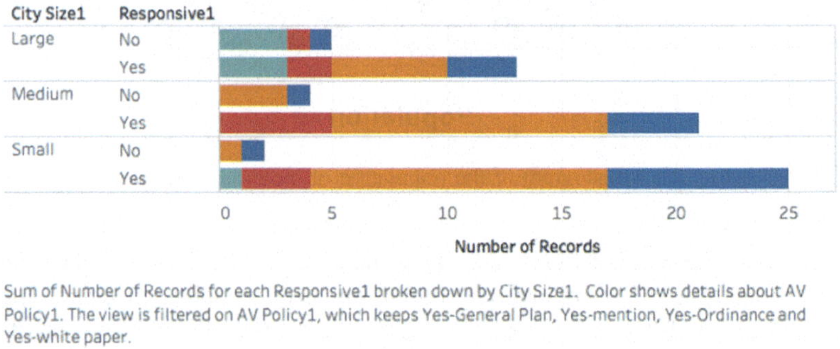

Fig. 4. Number of cities with AV policy with a responsive website by city size.

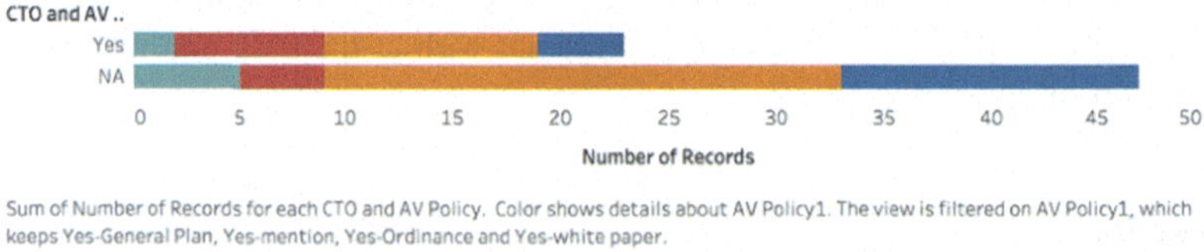

Fig. 5. Number of cities with AV policy with a CTO by type.

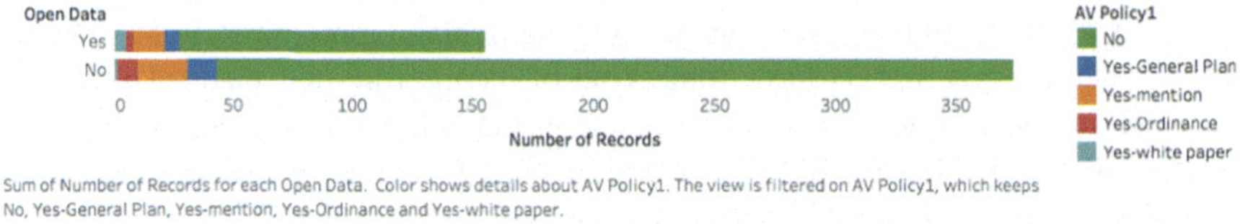

Fig. 6. Number of cities with AV policy with an open data initiative by type.

At the same time, as shown in Fig. 7, many of these cities are concentrated in the West and the South. Particularly when looking at actual policy (ordinances and general planning documents) the West and South far outpace the rest of the country. While reasons for this may be practical in that these locations are home to some of the companies doing the most testing of AVs currently, it is important to note this, since these locations may hold models and lessons for other locations moving forward.

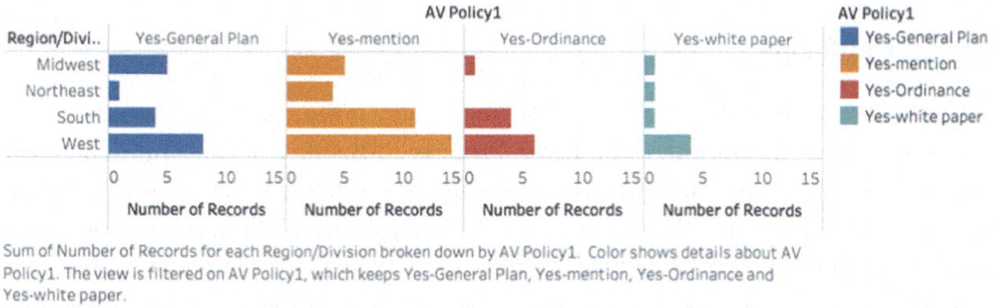

Fig. 7. Number of cities with AV policy by type and location.

With regard to the specific content of this these policies, most had general guidance. This focused on vehicular safety and operational conditions at the local level—something that is increasingly being preempted by state and federal policy. Examples of general language from a handful of cities are provided below.

4.1 San Antonio, Texas

NHTSA released a policy statement providing guidance in regards to testing automated vehicles [42]. The statement recommends for testing of self-driving vehicles at this time. This consideration may be lifted in the future as the technology progresses. If legislation and accompanying regulations to allow for the testing of AVs is pursued NHTSA recommends the establishment of a licensing program, on-road testing of self-driving vehicles, limited testing operations to roadway, traffic and environmental conditions, establish reporting requirements to monitor the performance of AV technology during testing. Other recommendations: (a) Ensure that the process from self-driving mode to driver control is safe (b) AVs should have the capability to detect, record, and inform the driver (c) Ensure the installation and operation of AVs does not disable any federally required safety feature or systems (d) Ensure AVs record information about the status of the automated control technologies in the even of a crash or loss of vehicle control.

4.2 Boston, Massachusetts

Before testing on streets companies must important standards including ease of manual takeover from autonomous mode, emergency braking and emergency stop functionality and basic driving capabilities such as staying with a lane [43]. Testing only occurs during optimal weather conditions and during daylight hours. Once a company reaches a certain milestone some limitations are lifted and testing can be done in other areas of Boston, at night-time, and during inclement weather. Testing in Boston includes the use of safety driver focused on roadway activity, as well as engineer monitoring the vehicles software. Companies must provide a history of their testing practices, documentation of extensive off-street and previous on-street testing, compliance with federal safety guidelines for autonomous vehicles and details of safety driver training procedures.

4.3 Chico, California

In Chico provides must have: "$5 million insurance, bond, or self-insurance • Communication link with the remote operator. • Process to display or communicate vehicle owner or operator information to a law enforcement officer. • AV complies with all FMVSS and CVC Div. 12 (Equipment of Vehicles), or NHTSA has approved an exemption. • Meets the description of level 4 or 5 automated driving system. • Law enforcement interaction plan • Remote operators + training program • Passengers that are not employees/contractors will be notified what personal information, if any, may be collected and how it will be used. • Annual report of disengagements to the DMV. • Report collision resulting in damage of property, bodily injury, or death to DMV within 10 days. • No charging of a fee or receiving other compensation for providing a ride to members of the public" [44]

4.4 Port St. Lucie, Florida

The policy in Florida is being updated by the standards of four parts: vehicle performance guidelines; model state policy; extensions of NHTSA's current regulatory tools and possible new regulatory actions [45]. In respect to model state policy and administrative tasks there should be a designation of a lead agency and form an autonomous technology committee with laws and regulations in regard to licensing/registration, driver education and training, insurance and liability, enforcement of traffic laws and regulations. Some policy considerations include the promotion of safety for pedestrians, bicyclists, transit rider and others, incentivized shared automated, electric vehicles to reduce the environmental impacts, and re-balance the use of the right of way with less space for cars and more space for pedestrians and cyclists.

4.5 Menifee City, California

Policies for requirements of driverless testing include certify local authorities have been provided written notification, submit a copy of a law enforcement interaction plan, maintain a program for remote operations and certify each operator has completed training and inform the DMV of the intended operation design domains [46].

4.6 Cities with Specific Suggestions: Seattle, Portland and San Jose

A handful of other cities offered specific suggestions. These were largely limited to the themes of management and design, and concentrated among cities in the West.

For example in Seattle preliminary automated mobility policy framework [47] aims to: (1) continue prioritizing the needs of people walking, biking, and taking transit and leveraging the growth of our robust transit network; (2) Support the development and testing of automated mobility technology, learning from the pilots and partnerships with local and national technology and operating equipment manufacturers; (3) Establish clear policy parameters that ensure automated vehicles help achieve the Mayor's five core values and out shared and emerging mobility principles - not counteract them. Goals include:

- Enacting a "people and transit first" approach to AVs.
- Allow a mix of fleet, combination of human-driven and fully automated vehicle operations within the City of Seattle to eliminate the dangers or partial automation.
- Encode base operating parameters into connected and automated vehicles (i.e. speed limits).
- Collaboration with federal and state policymakers to ensure SDOT's core local controls and police powers related to AV regulations are not preempted.
- Establish time-based access restrictions or pricing for geofenced congestion management Requirement of detailed data from all vehicles.
- Protect the privacy of individuals by anonymizing personally identifiable data generated by connected and automated vehicles.

Policies also include regulations in equity and accessibility, pilots and partnerships, and infrastructure and street design, mobility economics, and land use and building design.

Likewise in Portland, policy ensures that all levels of AV are operating safely for all users, requiring the adequate insurance coverage for operators, customers, and the public at-large by providers of connected and AVs [48]. Also that the connected and automated vehicles improve travel time reliability and system efficiency by (1) maintaining or reducing the number of vehicle trips during peak congestions (2) reduce low occupancy vehicles (3) pay for use of and impact on transportation systems including factors such as congestion and vehicle miles traveled, vehicle occupancy, and vehicle energy efficiency. (4) Support and encourage the use of public transportation. Overall effect of lowering carbon pollution by reducing low occupancy vehicles, as well as make the benefit of automated mobility available on an equitable basis to all segments of the community. In the end identify, prevent, and mitigate potential adverse impacts from connected and automated vehicles.

> *Portland, Connected and Autonomous Vehicles Policy, (2017)*
> *Policy 9.xx Connected and Autonomous Vehicles. Ensure that connected and autonomous vehicles advance Portland's Comprehensive Plan multiple transportation goals and policies, including vision zero, climate pollution reduction and cleaner air, equity, physical activity, economic opportunity, great places, cost effectiveness, mode share, and reducing vehicle mile traveled.*

Finally, the City of San Jose continues the trend toward more broad and integrated approaches to integrating AVs into transportation policy [49]. These more thematic policy statements in the arena of management and design focus on things like transit systems; parking; curbside management and data systems. As articulate in the Smart City Vision for San Jose, the city wants to be a "demonstration city."

> *Demonstration City: Reimagine the City as a laboratory and platform for the most impactful, transformative technologies that will shape how we live and work in the future.*
> *Fully develop the city's transportation innovation zone to test new products and services, such as autonomous vehicles, that will dramatically shape transportation in the future and mitigate traffic congestion.*
> *Build an "Internet of Things" platform employing transit vehicles and infrastructure by using smart sensor technologies to improve safety, mobility, and optimize our transit system.*
> *Create pathways for start-ups and innovators to easily access opportunities to pilot and test new products and services with the City, such as by hosting "demo days" to highlight the most innovative "smart city" companies in Silicon Valley, and sponsoring public competitions to encourage crowdsourcing of innovative solutions to civic challenges.*

Some documents in these cities, for example Portland, referenced complete street policy and placemaking strategy. Virtually all focused on energy and electric vehicle infrastructure. While there was some mention of creating "pathways for innovation" to pilot and test new mobility (for example in the City of San Jose's policy), there was little discussion about ways to speed the policy process or increase the speed of decision making in an era of new mobility. Further noticeably absent were dialogues about travel behavior programs or how traditional transportation demand management (TDM) may need to adapt or evolve.

Many expert planners and engineers surveyed (N = 72) at AVS echoed this sentiment. As is shown in Fig. 8 while many optimistically predict growth in both transit ridership and multimodal travel an equal amount predict that there will be either no change or a decline in these modes of transport as autonomous vehicles come to fruition, yet at the same time when asked to articulate their greatest hopes or fears in 3 words or less, many expressed concerns about the direction of policy to achieve safe, green and avoid congested, inequitable streets.

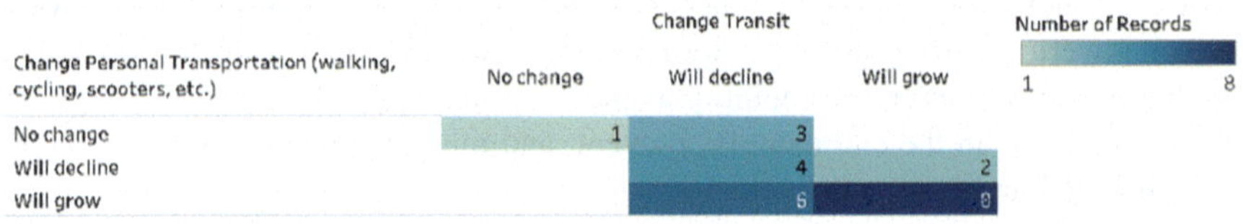

Fig. 8. Expert panel on changes to transit and non-automotive modes.

As shown in Fig. 9, top desired themes (or "hopes") included: safe/very safe; efficient; green; people/for people/people-first; quiet; livable/lively; accessible; friendly; shared. Top "fears" were that streets would become: congested; unsafe; unwalkable; inequitable.

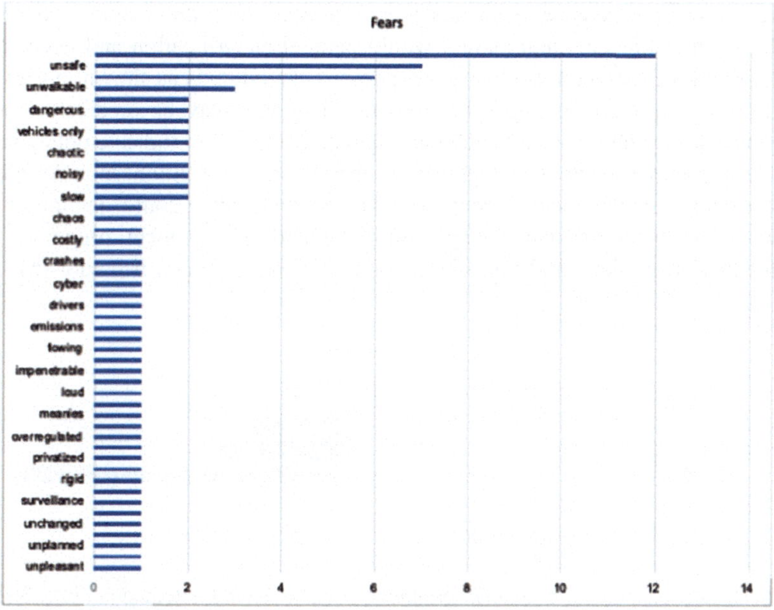

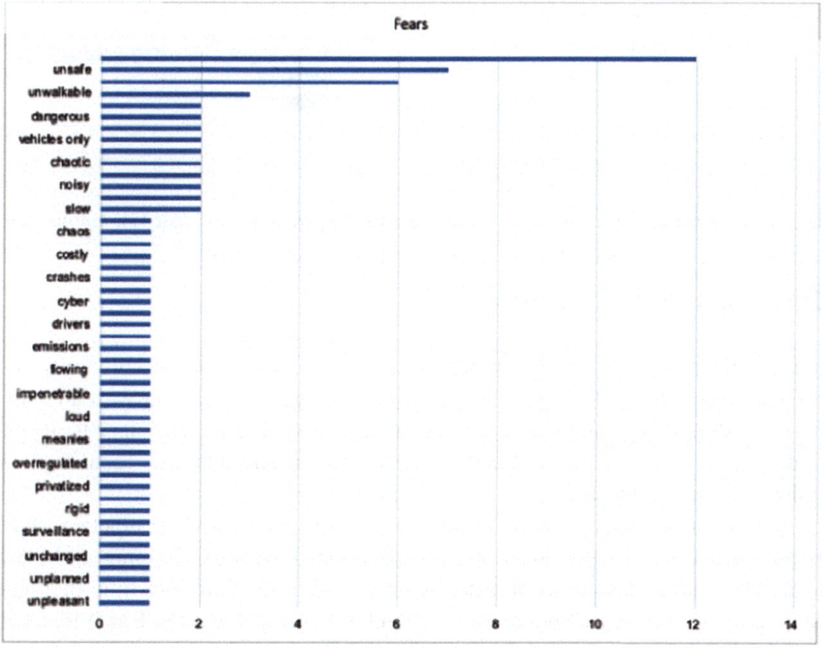

Fig. 9. Expert panel hopes and fears.

At the same time most believed that there was time to address policy concerns. Most believed that full-fleet automation would parallel electrification and occur after 2040, with the most common use being autonomous shared taxis as shown in Figs. 10 and 11. This is important to highlight since much of this may be predicated on an assumption of policy action—yet benchmarking data continues to indicate policy lag—particularly in policy aspects that might drive shared use. These might include pricing and right-of-way management incentives [34, 50] but also could include more aggressive travel behavior and transportation demand management programs that harness the power of social and financial markets to "nudge" travel behavior [51–54].

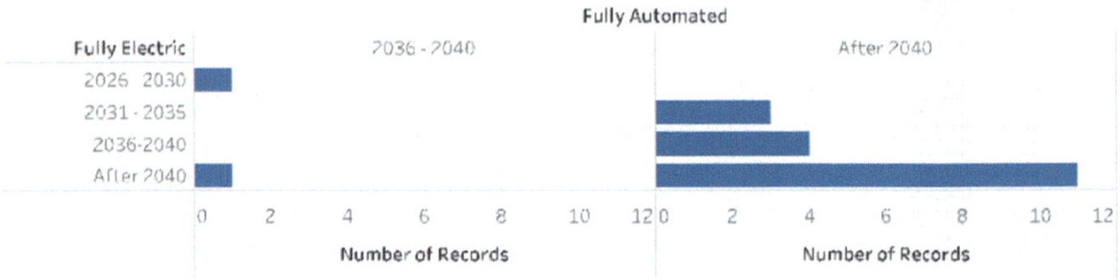

Fig. 10. Expert panel view on market proliferation skews to full automation post 2040.

Fig. 11. Expert panel view on key use of autonomous vehicles by year full automation.

5 Discussion and Conclusion

In sum an increasing yet small number of cities are thinking about autonomy—with only 5% having actual policy as of 2018. This will be an important growth area, in the future, yet it should be noted that it is the content that should be a primary focus going forward—along with making that content transactional and adaptive (consistent with generalized theory in the area).

At the same time when looking at this policy there are some best practices that are emergent, particularly in the areas of curbside policy, roadway thinning and parking removal. There also appears to be a focus on energy transition. Yet little dialogue is focused on the processing of policy making and governing and even less if focused on travel behavior.

This is surprising given the context of increase e-governance and changing travel behaviors. There are numerous emergent tools that allow for more agile and nimble policy dialogues, whether they be those that offer visualization (e.g. ReStreet, Remix, etc.), those that offer data dashboard (MySidewalk, Populus) or those that provide for a more seamless way to engage the public (Mindmixer, etc.). These kind of tools are of increasing importance as mobility continues to be disrupted and cities fail to adapt quickly enough.

For example, the integration of e-bikes and e-scooters into cities has seemed almost instantaneous. Using some of the same technology as the Segway, companies such as Bird, Lime, and Spin are offer first- and last-mile solutions on e-bikes and have millions of dollars in funding to expand. According to *Wired* magazine the company "gave more than 95,000 rides to 32,000 different people in just its first 30 days of service in the city" [55]. According to Bird their scooters provide an "unprecedented opportunity to reduce car trips… roughly 40% of trips under two miles—thereby reducing traffic, congestion, and greenhouse gas emissions" [56]. In places like San Francisco, there was backlash over concerns such as scootering on the sidewalk and helmets along with the notion of how they may create on the streets for pedestrians and cyclist, that led to a slow-down and eventual elimination of the scooters for a time [57]. The inability to evolve policy fast enough became an impediment to sustainable transport.

Further, specifically focused on travel behavior, little is being done to integrate or rethink TDM and travel programs. There are also many cities that are thinking about policies to induce behavioral change, be they congestion pricing or kinds of incentives, but this has yet to include dialogue on AVs. How could policy on AVs advance these programs? Could it harness innovation in incentive-based and behavioral programs? Opportunities might be found in new ways of providing affordable or free transit, capitalizing on social norms and gifts, making greater use of targeted marketing and harnessing data, information or emotional pleas. While exploring these norms have been discussed in some TDM literature [51], they may offer a new possibility as AVs become more prevalent and they will need to become a focus as driving becomes easier and more frictionless.

References

1. Bahamonde Birke, F.J., Kickhöfer, B., Heinrichs, D., Kuhnimhof, T.: A systemic view on autonomous vehicles: policy aspects for a sustainable transportation planning (2016)
2. Fagnant, D.J., Kockelman, K.M.: The travel and environmental implications of shared autonomous vehicles, using agent-based model scenarios. Transp. Res. Part C Emerg. Technol. **40**, 1–13 (2014)
3. Schlossberg, M., Riggs, W.W., Millard-Ball, A., Shay, E.: Rethinking the street in an era of driverless cars. UrbanismNext (2018). https://urbanismnext.uoregon.edu/files/2018/01/Rethinking_Streets_AVs_012618-27hcyr6.pdf
4. Sperling, D.: Three Revolutions: Steering Automated, Shared, and Electric Vehicles to a Better Future. Island Press (2018)
5. Riggs, W., Boswell, M.R.: Thinking Beyond the (Autonomous) Vehicle: The Promise of Saved Lives (2016)
6. Riggs, W.W., Boswell, M.R.: "No Business as Usual in an Autonomous Vehicle Future," 2016
7. Riggs, W.: Disruptive Transport: Driverless Cars, Transport Innovation and the Sustainable City of Tomorrow. Routledge, London (2019)

8. Guerra, E.: Planning for cars that drive themselves metropolitan planning organizations, regional transportation plans, and autonomous vehicles. J. Plan. Educ. Res. 0739456X15613591 2015. https://doi.org/10.1177/0739456x15613591
9. Riggs, W., Steins, C., Chavan, A.: City Planning Department Technology Benchmarking Survey 2015. Planetizen: The Urban Planning, Design, and Development Network (2015). http://www.planetizen.com/node/73480/city-planning-department-technology-benchmarking-survey-2015. Accessed 20 May 2015
10. Cabral, J.E., Chavan, A., Clarke, T.M., Greacen, J.: Using technology to enhance access to justice. Harv. J. Law Technol. **26**, 241 (2012)
11. Chavan, A.: Best practice benchmarking for legal services websites 2014| urban insight. Urban Insight, 24 April 2014. https://www.urbaninsight.com/2014/04/24/best-practice-benchmarking-for-legal-services-websites-2014. Accessed 30 Sept 2014
12. Riggs, W.: Mobile responsive websites and local planning departments in the US: opportunities for the future. Environ. Plan. B Plan. Des. p. 0265813516656375, July 2016, https://doi.org/10.1177/0265813516656375
13. Evans-Cowley, J., Manta Conroy, M.: The growth of e-government in municipal planning. J. Urban Technol. **13**(1), 81–107 (2006), https://doi.org/10.1080/10630730600752892
14. Evans-Cowley, J.: Planning in the real-time city: the future of mobile technology. J. Plan. Lit. **25**(2), 136–149 (2010). https://doi.org/10.1177/0885412210394100
15. Kwak, H., Lee, C., Park, H., Moon, S.: What is Twitter, a social network or a news media? In: Proceedings of the 19th International Conference on World Wide Web, pp. 591–600 (2010)
16. Ling, R.: The Mobile Connection: The Cell Phone's Impact on Society. Morgan Kaufmann, San Francisco (2004)
17. Riggs, W., Gordon, K.: How is mobile technology changing city planning? Developing a taxonomy for the future. Environ. Plan. B Plan. Des. 0265813515610337 (2015), https://doi.org/10.1177/0265813515610337
18. Slotterback, C.S.: Planners' perspectives on using technology in participatory processes. Environ. Plan. B Plan. Des. **38**(3), 468–485 (2011). https://doi.org/10.1068/b36138
19. Evans-Cowley, J., Kubinski, B.: A brave new world: how apps are changing planning. Planetizen: The Urban Planning, Design, and Development Network, September 2012. http://www.planetizen.com/node/58314. Accessed 15 Dec 2014
20. Guerra, E.: When autonomous cars take to the road. Planning, **81**(5) (2015)
21. Larco, N.: When Are AVs Coming? (10 Car Companies Say Within the Next 5 Years…)| Urbanism Next (2017). https://urbanismnext.uoregon.edu/2017/08/28/when-are-avs-coming-10-car-companies-say-within-the-next-5-years/. Accessed 29 September 2017
22. Clewlow, R.R., Mishra, G.S.:Disruptive Transportation: The Adoption, Utilization, and Impacts of Ride-Hailing in the United States," University of California, Davis, Institute of Transportation Studies, Davis, CA, Research Report UCD-ITS-RR-17-07 (2017)
23. Clark, B., Larco, N.: The Impacts of Autonomous Vehicles and E-commerce on Local Government Budgeting and Finance. UrbanismNext (2018)
24. Rayle, L., Dai, D., Chan, N., Cervero, R., Shaheen, S.: Just a better taxi? A survey-based comparison of taxis, transit, and ridesourcing services in San Francisco. Transp. Policy **45**, 168–178 (2016). https://doi.org/10.1016/j.tranpol.2015.10.004
25. Shaheen, S., Chan, N.: Mobility and the sharing economy: potential to facilitate the first-and last-mile public transit connections. Built. Environ. **42**(4), 573–588 (2016)
26. Rassman, C.L.: Regulating rideshare without stifling innovation: examining the drivers, the insurance gap, and why Pennsylvania should get on board. Pitt. J. Tech. Pol. **15**, 81 (2014)
27. Riggs, W.W., Boswell, M.R., Zoepf, S.: A New Policy Agenda for Autonomous Vehicles: It's Time to Lead Innovation, Planetizen (2017)

28. Lipson, H., Kurman, M.: Driverless: Intelligent Cars and the Road Ahead. MIT Press, Cambridge (2016)
29. Harb, M., Xiao, Y., Circella, G., Mokhtarian, P., Walker, J.: Projecting travelers into a world of self-driving cars: naturalistic experiment for travel behavior implications. In: Proceedings of the 97th Transportation Research Board, Washington D.C., (2018)
30. Appleyard, B., Riggs, W.: 'Doing the Right Things' Before "Doing Things Right": a conceptual transportation/land use framework for livability, sustainability, and equity in the era of autonomous vehicles. In: Presented at the transportation research board 97th annual meeting transportation research board, Washington, D.C. (2018)
31. Riggs, W.: Technology, civic engagement and street science: hacking the future of participatory street design in the era of self-driving cars. In: Proceedings of the 19th Annual International Conference on Digital Government Research: Governance in the Data Age, Delft, Netherlands, pp. 4:1–4:6 (2018). https://doi.org/10.1145/3209281.3209383
32. Riggs, W., Appleyard, B., Johnson, M.: A design framework for livable streets in the era of autonomous vehicles. In: Proceedings of the 98th Annual Meeting Transportation Research Board, Washington, D.C. (2019)
33. Riggs, W., Larco, N., Tierney, G., Ruhl, M., Karlin-Resnick, J., Rodier, C.: Autonomous vehicles and the built environment: exploring the impacts on different urban contexts. In: Meyer, G., Beiker, S. (eds.) Road Vehicle Automation 5, pp. 221–232. Springer, Cham (2019)
34. Crute, J., Riggs, W., Chapin, T., Stevens, L.: Planning for Autonomous Mobility. American Planning Association, Washington D.C., PAS 592 (2018)
35. NACTO: Blueprint for Autonomous Urbanism. National Association of City Transportation Officials, New York (2017)
36. Bloomberg: Bloomberg Aspen Initiative on Cities and Autonomous Vehicles (2016). https://avsincities.bloomberg.org/. Accessed 19 Mar 2018
37. McKinsey: Automotive revolution – perspective towards 2030 (2016)
38. Hoene, C.W., Pagano, M.A.: Research Brief on America's Cities. National League of Cities, Washington, D.C. (2013)
39. Brudney, J.L., Selden, S.C.: The Adoption of innovation by smaller local governments: the case of computer technology. Am. Rev. Public Adm. **25**(1), 71–86 (1995). https://doi.org/10.1177/027507409502500105
40. Van der Meer, A., Van Winden, W.: E-governance in cities: a comparison of urban information and communication technology policies. Reg. Stud. **37**(4), 407–419 (2003). https://doi.org/10.1080/0034340032000074433
41. Vogelsang, I., Compaine, B.M.: The Internet Upheaval: Raising Questions, Seeking Answers in Communications Policy. MIT Press, Cambridge (2000)
42. City of San Antonio, "AV Policy," City of San Antonio (2018). https://library.ctr.utexas.edu/ctr-publications/0-6803-p2.pdf. Accessed 20 Jan 2020
43. City of Boston: AV Approach (2018). https://www.boston.gov/departments/new-urban-mechanics/autonomous-vehicles-bostons-approach. Accessed 20 Jan 2020
44. CA DMV: California DMV Presentation (2018). http://www.dot.state.mn.us/automated/docs/meetings/advisory/july2018/cadmvpresentation.pdf
45. Florida APA: AV Guidance, Florida Planning. http://www.floridaplanning.org/wp-content/uploads/2018/05/MPO-Guidance-for-Automated-Transit-Final-Report.pdf
46. CA DMV: CA DMV News Release (2018). https://www.dmv.ca.gov/portal/dmv/detail/pubs/newsrel/2018/2018_17
47. City of Seattle: New Mobility Program (2018)
48. City of Portland: Connected and Autonomous Vehicle Policy. PBOT SAVI Program (2017). https://www.portlandoregon.gov/transportation/article/643814

49. City of San Jose: Smart City Vision (2018)
50. Bianco, M.J.: Effective transportation demand management: combining parking pricing, transit incentives, and transportation management in a commercial district of Portland, Oregon. Transp. Res. Rec. J. Transp. Res. Board **1711**(1), 46–54 (2000)
51. Riggs, W.: Painting the fence: Social norms as economic incentives to non-automotive travel behavior. Travel Behav. Soc. **7**, 26–33 (2017). https://doi.org/10.1016/j.tbs.2016.11.004
52. Riggs, W., Kuo, J.: The impact of targeted outreach for parking mitigation on the UC Berkeley campus. Case Stud. Transp. Policy **3**(2), 151–158 (2015). https://doi.org/10.1016/j.cstp.2015.01.004
53. Riggs, W.: Testing personalized outreach as an effective TDM measure. Transp. Res. Part Policy Pract. **78**, 178–186 (2015). https://doi.org/10.1016/j.tra.2015.05.012
54. Weber, J., Azad, M., Riggs, W., Cherry, C.R.: The convergence of smartphone apps, gamification and competition to increase cycling. Transp. Res. Part F Traffic Psychol. Behav. **56**, 333–343 (2018). https://doi.org/10.1016/j.trf.2018.04.025
55. Aarian, M.: The love of the people isn't enough to keep shared electric scooters rolling. Wired Mag. (2018). http://www.wired.com/story/shared-electric-scooters-rolling/http://www.wired.com/story/shared-electric-scooters-rolling/.
56. Bird: Bird (2018). http://www.bird.co/
57. Inman: Real estate agents see electric scooter boom as mixed blessing. WFG National Title Insurance Company, 18 May 2018. https://national.wfgnationaltitle.com/2018/05/18/real-estate-agents-see-electric-scooter-boom-as-mixed-blessing/. Accessed 07 July 2018

Ethical Algorithms in Autonomous Vehicles: Reflections on a Workshop

Nicholas G. Evans(✉)

Department of Philosophy, University of Massachusetts Lowell, Lowell, USA
Nicholas_Evans@uml.edu

Abstract. This chapter summarizes and expands on the breakout session at the 2019 Autonomous Vehicles Symposium in Orlando, FL, titled "Ethical Algorithms in Autonomous Vehicles." First, the content of the workshop presentations is summarized, covering technical and nontechnical detail. Second, the discussion during, and on the margins of the workshop is summarized from the perspective of the authors, consistent with the Chatham House Rule. Conclusions are posed for industry, academia, and government readers.

Keywords: Autonomous vehicles · Ethics · Crash algorithms · Safety data · Algorithmic bias · Risk management

1 Introduction

This chapter summarizes and expands on the breakout session at the 2019 Autonomous Vehicles Symposium (AVS) in Orlando, FL, titled "Ethical Algorithms in Autonomous Vehicles." This workshop-styled session was funded through the National Science Foundation grant by the same name, and was centered around ethical issues arising from the development and implementation of autonomous vehicles (AVs). The workshop took place over two days of the AVS, Monday 15th and Tuesday 16th of July.

The primary aim of the workshop was to bring together internationally recognized and emerging scholars in ethics and policy to present new work in machine ethics and the ethics of AVs. The two days of programming featured discussion and conceptual innovation in the ethics of autonomous vehicles, including an open forum to identify emerging issues and develop collaborations for future work. Presenters were drawn from diverse career levels; and from both private and academic enterprise.

The central motivation for such a workshop was two-fold. First, although considerable attention has been paid to the basics of ethics in AVs, very little work has been done to determine a) what specific ethical theories say about machine behaviors; and b) what kinds of behaviors and traits of AVs are of greatest concern. To date, a large proportion of work has focused on surveys of consumer preferences (Bonnefon et al. 2016; Awad et al. 2018), or preliminary work establishing the case *that* there is something ethically important about autonomous vehicles (Goodall 2014a, b, 2016a, b; Lin 2015). The aim of this workshop was to extend this analysis, and that of the

G. Meyer and S. Beiker (Eds.): AVS 2019, LNMOB, pp. 161–169, 2020.
https://doi.org/10.1007/978-3-030-52840-9_15

previous year's breakout session (Goodall et al. 2018), and allow for a frank discussion of ethical challenges facing autonomous vehicles, informed by the expert community AVS attracts.

While individual papers and their content are available on the AVS 2019 website, discussion took place under the Chatham House Rule (CHR). CHR is a common tool in security and diplomatic discussions, in which comments from participants may be quoted, but may not be attributed or identified in any way. This has a marked advantage over closed workshops in which information may be spoken freely, but never disclosed, by providing nonparticipants a record of the conversation that occurred. However, limitations arise because unless authors choose to personally identify themselves and their comments, it is difficult to vet information.

This chapter proceeds as follows, and with the following methodological turn. In the next section, we survey the conference proceedings, covering the information relayed by the invited speakers at the workshop. We then turn to the discussion at, or on, the margins of the workshop, and thoughts by speakers and participants on future work in the ethics of AVs. We conclude with some recommendations for those working in industry, academia, and government.

2 Workshop Content

The workshop was broken into two parts, covering two full breakout sessions of content. Each day began with opening remarks and logistics by the conference organizers, followed by a series of talks, each with time allocated for open discussion.

The first session consisted of a keynote describing preliminary research into modelling ethical decisions for autonomous vehicles using naturalistic crash data. This presentation, by Nicholas G. Evans and Rocco Casagrande, argued that consumer preference models of autonomous vehicles, including the MIT Moral Machine project, fail to give accurate insights into ethical issues on two counts. The first count being that these projects typically stipulate pairwise comparisons between two scenarios, without considering a full range of options, or exploring what options might exist for AVs. Secondly, consumer preferences fail to give ethical guidance because they simply display preferences, rather than considered reasons for what autonomous vehicles should do. The presentation concluded with an examination of an empirical case, showing how parametric models could be used to explore a full range of ethical features for important individual cases.

Katherine Evans then spoke about if, and how, the AVs should be limited in the kind of information they have on hand. With the arrival of 5G technology, as well as the increased efficiency of vehicle-to-vehicle communication, vehicle-to-device communication, and the Internet of Things, looms a second ethical question: what morality requires an AV to know about its decision context, and—perhaps even more importantly—what an AV should not know about the users in its environment. Embracing a liberal view on data protection laws and technical capability, the future of AV decisions could look more like a real-life rendition of the Moral Machine Experiment; a world where AVs may be able to identify the old, the rich, and the criminals in their midst, and incorporate these features into their moral deliberation. Other features may also seem topically salient: the health status of different pedestrians, the credit score of a driver, or even their tacit and explicit social and political affiliations. The collection and exchange of user data across different devices could afford an AV a quasi-omniscient perspective from which to make moral decisions, but it is a separate question to know which types of information, if any, should make a moral difference.

A panel discussion then ensued examining particular challenges in the ethics and policy of designing navigation algorithms for AVs. First, Duncan Purves explored whether the asymmetry in public opinion between "autonomous weapons" and AVs is coherent. He articulated some obvious and subtle reasons for the asymmetry by way of addressing several recent arguments for banning autonomous weapons systems, advanced by some academics and NGOs, and found that—perhaps surprisingly—some of these objections to autonomous weapons systems also seem to apply to AVs. He then suggested that the difference in public attitudes about AVs and autonomous weapons systems is perhaps best explained by our feelings about the morality of the larger enterprises in which they are deployed: transportation and war. This compels us to justify the larger transportation system, of which AVs are only a recent innovation.

Next, Damien Williams tackling the problem of the dominance of Western ethical perspectives in the development of AVs. In considering the question of what AVs ought to do, Williams argued, designers, coders, and trainers will need to develop new ways of training and categorizing the decision-making processes of the algorithmic systems at work in AVs to account for the cultural and moral concerns of nonwestern societies. By including different nonwestern understandings of nonhuman agency from Asian and West African societies, we can explore new ways of thinking about assemblages of human and machine action, toward the maintenance and enrichment of human and nonhuman life. With this done, we can train AVS to make decisions in a wide variety of global cultural contexts, to address the needs and concerns a wide range of stakeholders.

Finally, Sarah Thornton showcased the Designing for Human Values (DHV) framework including a set of concepts, methodologies and tools for addressing ethical considerations throughout the engineering of a technology. DHV has its roots in Value Sensitive Design (VSD), which is an open-ended design framework that helps to analyze technology in terms of the human values that technology expresses. As a framework, VSD prompts the designer to focus on a broad set of stakeholders impacted by the technology under consideration (e.g. users, policy makers, the environment, the public), the values attached to those stakeholders (e.g. privacy, trust, profit), and the value tensions that can arise between different competing values and associated stakeholders. DHV adds structure to the underlying VSD framework so engineers can incorporate ethical considerations efficiently, reliably and consistently when designing a technology: exposing relevant ethical issues related to a particular technology upstream in the engineering design process; and prompting engineers to identify and reason through design options in more detail, and with a more informed, nuanced and critical eye, than they would have otherwise. She presented results from several 3-h DHV Workshops with diverse teams from industry and academia in order to demonstrate the use of the DHV framework.

Geoff Keeling then presented ongoing work on the kinds of confidence AVs ought to have in certain objects. First, AVs are morally required to exercise *due caution* around vulnerable road-users such as pedestrians and cyclists. Presumably, this amounts to reducing speed and performing maneuvers to lower the risk of colliding with the road-user within some morally acceptable range. Second, this same degree of caution is not, in general, required towards road-users *inside* other vehicles. The asymmetry in the amount of caution required is explained by the fact that the expected harm to vulnerable road-users is significantly greater than the expected harm to non-vulnerable road-users in collisions with similar impact velocities. Hence the true class for an object, e.g. pedestrian, is morally relevant insofar as the morally right level of caution for the AV to exercise towards an object depends on what kind of thing the object is. Third, in scenarios like the Tempe collision, where the AV is uncertain about the classification for an object, the morally right level of caution is also uncertain. My aim in this paper is to investigate the degree of caution which AVs are morally required to exercise in scenarios like these. The view that I defend is deontological. Roughly, my thesis is that the AV should behave *as if* an object is a vulnerable road-user just in case it is reasonable for an epistemic agent with the same evidence as the AV to believe that the object is a vulnerable road-user. Conversely, the AV is permitted to behave *as if* an object is a non-vulnerable road-user just in case it is reasonable for an epistemic agent with the same evidence as the AV to believe that the object is a non-vulnerable road-user. I spell-out the meaning of these conditions with reference to the AV's probability distribution over the different classes which an object might belong to.

Kendra Chilson presented on the issue of consumer trust, and its importance in the development of autonomous vehicles (A.V.s). Without this trust, A.V.s would not only under-perform, they could be unusable and dangerous. Chilson gave an epistemic account of trustworthiness that beyond the conditions for manufacturers to cultivate consumers' subjective trust. Instead, her account identified a "robust trustworthiness," which fully justifies consumers' trust in A.V.s, based on appropriate indicators of trustworthiness. She developed six desiderata for autonomous systems, based on analogy to automatic technologies:

1) Repeatability—whether the system can be put back into the same state and produce the same outcome
2) Predictability—whether an expert can determine, based on input, what the system will output
3) Reliability—whether anyone can depend on the system's behavior by forming expectations about its behavior through repeated interactions
4) Transparency—whether an expert can understand what is happening within the system in real time
5) Re-constructability—whether the system's trajectory toward producing a certain output can be reconstructed after the fact (even if that trajectory is non-repeatable)
6) Explicability—whether the system can give a high-level explanation of what it is doing and why that is accessible to anyone

She then argued that these concepts can be divided along two orthogonal axes: System Externality vs. Internality, and Expert vs. Non-expert groupings. I will show why, to establish appropriate trust for non-expert consumers, both external and internal non-expert indicators must be present, and internal expert indicators must be accessible by appropriate authorities. Then through conceptual analysis, I will show that several of these desiderata entail the other indicators needed to establish robust trustworthiness for A.V.s (given certain background assumptions), and give recommendations for incorporating these features into A.V.s.

The final panel concerned the wider effects of AVs on society. William Bauer argued that the widespread introduction of AVs presents macro-level socioeconomic concerns in addition to micro-level 'ethics on the road' dilemmas. His paper examined the impacts of AVs on our basic socioeconomic structure. Given the advent of self-driving freight trucks and taxis, for instance, millions of drivers could become unemployed very quickly. Citizens and leaders should address these kinds of macro-level social justice issues in advance in order to forestall the worst outcomes. To plausibly address these issues, he and his coauthor applied the principles of John Rawls' theory of distributive justice to the question of AVs. Focusing on the cases of truck and taxi drivers, we argue that the principles of Justice as Fairness support several possible policy-guiding norms that can be used to develop specific regulatory policies and ensure smoother economic transition as AVs are implemented on a broad scale.

Next, Johannas Himmelreich examined passenger settings for AVs. He identified conflicts between values in AV programming, and argued that passengers should be allowed to set the parameters to solve such value conflicts. Importantly, however, the parameter setting must be interdependent and passengers should not be allowed to solve conflicts independently of each other. Figuratively speaking, a passenger should have only one control dial to solve the value conflicts instead of multiple dials. The two conflicts Himmelreich identified were between mobility (e.g. time expected to arrive at a destination) and safety (e.g. route planning and speed control in navigating around obstacles); and between passenger-interests and outsider-interests, including collision management and speed control for passenger comfort. He provides an ethical analysis of these four values in conflict and draw on basic microeconomics to formalize the conflict. He defended a dependent passenger parameter setting because this promotes the meta-values of pluralism, human agency, and fairness.

Finally, Carole Turley Voulgaris examined the potential for connected and autonomous vehicles' (CAVs') potential to improve users' quality of life by reducing the frustration and inefficiency associated with traffic congestion. Traffic congestion is a function of the ratio of the number of vehicles using a roadway (volume) and the maximum number of vehicles that the roadway can accommodate (capacity). Vehicle connectivity and autonomy could indeed reduce congestion by enabling fleets of vehicles to coordinate their movements more efficiently, thereby increasing the effective capacity of a roadway. However, since CAV users—freed from the task of vehicle operation— could use their travel time for more pleasant or productive activities, automation would also increase travelers' tolerance for traffic congestion, increasing the demand for motorized travel and likely returning congestion to (and even beyond) levels experienced prior to the introduction of CAVs. The negative effects of vehicular congestion extend beyond vehicle users' lost time to other harms shared with non-users, such as pollution exposure, climate change, and hostile land development patterns. By increasing travelers' tolerance for congestion, CAVs have the potential to shift the burden of congestion-related harms from vehicle users to non-users. Since vehicle ownership is highly correlated with income —and this relationship may be even stronger for CAVs—this would represent a benefit to higher-income households at the expense of lower-income households. Modifications to vehicle routing algorithms, well-designed roadway user fees, proactive land-use planning, and policies to encourage vehicle sharing could facilitate a more just distribution of the benefits of CAVs.

3 Workshop Discussion

As a preliminary discussion, the appearance of the 2019 novel coronavirus outbreak disrupted the timeline for the drafting of this chapter due to commitments from some of the workshop panelists and conveners. As such, and on a reduced timeframe, this discussion is based primarily on the comprehensive minutes collected by the second author, combined with the reflections of the primary author.

"Before The Crash" Ethics
A primary concern raised by participants at the workshop was the continued focus of ethical analysis on crash algorithms, rather than what one participant described as "the ethics of the crash, before the crash." That is, rather than attempting to resolve tradeoffs that arise in emergent scenarios (including crashes), members of the audience were interested in whether algorithms could be trained to recognize potential crashes and navigate around those situations without having to resolve them at the point of catastrophe. In the case provided by Evans and Casagrande, involving a human tailgater, the ideal would be to resolve the tailgating quickly rather than wait until an emergent case occurred.

This is something that ethicists have begun to consider, as part of a broader attitude towards the ethics of risk in the setting of AVs (Goodall 2016b). Such a strategy presents novel ethical issues, such as the degree to which an AV could hold up traffic in an effort to slow down to minimize the chance of an injurious collision with the tailgater; or speed up, potentially past the speed limit, to get out of their way (e.g. Jiang et al. 2005). We might call this a problem of *vagueness* regarding our emergent cases: attempting to avoid the case leads to a novel set of risk management problems, which carry with them their own ethical issues.

Importantly, however, considering strategies to avoid these emergent cases does not mean we can ignore them. The primary reason for this is that some kind of emergent case will surely exist even in the most optimized system. In the case of the tailgater, we could provide an example of a networked set of cars (including, potentially, human driven cars) that obviates the need for the decisions described by Evans and Casagrande. At the same time, however, such a network may have cascading effects that arise from small misalignments in the decisions of vehicles—in the same way that stock trades between algorithms can cause sudden shocks in stock prices. This is a novel potential emergent even, albeit different from the example of the tailgater. The central point here is that while good engineering might eliminate some ethical dilemmas, they will almost certainly leave some open (and even introduce new ones).

Concrete Analysis of Ethical Problems

A second issue raised during the workshop concerned the lack of specificity about what constitutes and ethical theory about AV behavior. While "applied ethics" is full of research *describing* problems, it is famously unclear about what comprehensive moral views people ought to take about those problems. The rare exception to this is military ethics, in which the combination of rigorous moral analysis and the long history of International Humanitarian Law provide a basis to think about the ethics of risk and come up with robust, detailed conclusions.

Rather, what has arisen in the ethics of AVs are broad assessments of "issues" and a larger vacuum of critical analysis, high quality empirical ethics work, and decision-making tools. Each has their own place, but none of the work in either is responsive to any of the others. This is a significant methodological problem, as neither program of work—conceptual, empirical, or decision-making—has the tools, in isolation, to provide necessary guidance to OEMs and other stakeholders. Deep conceptual work, like the work done in military ethics, is needed to very precisely articulate what our obligations are to road users, pedestrians, and the public at large, when considering how we impose risk through the deployment of AVs. Empirical work is needed to supplement the elements of conceptual work that rely on evidence to motivate one conclusion or another. And decision-making tools are required to interpret both into schemes that engineers and other specialists can apply to their work, without requiring a PhD in some other field (or multiple fields).

Larger Context

A tension proceeded as the workshop went on, between participants who favored discussion of ethics and navigation/risk management algorithms used to pilot AVs, and those who favored discussion of the wider context of AVs in society. This tension is reflected in the current ethics literature, which while overwhelming favoring the former acknowledges the latter as in need of urgent debate. The workshop brought to light some of these emerging issues, including the impact of AVs on congestion, on labor rights, and on cultural sentiments.

The need for larger debate, however, entails a need for a broader set of participants at AVS and elsewhere. In particular, ethical issues pertaining to broad social effects are typically the domain of political philosophers, who consider the basic structure of social institutions as part of their work. The policy conversation around these issues, moreover, requires policy*makers* from agencies such as the Department of Labor, Department of Commerce, and others. Expanding the sphere of concern around AVs requires inclusion of a greater representation of stakeholders. This carries logistical burdens, but would also create new opportunities and benefits to design AVs for society—and better design society for AVs.

Resources and Funding

As a final note, considerable diversity was added to the breakout session over previous years, with the inclusion of emerging scholars in the field of applied ethics, and novel methodologies transplanted from other disciplines into the subject of AV ethics. Continuing this trajectory was judged to be desirable by attendees of the workshop. However, without continued funding, the cost of producing these breakout sessions—while modest in absolute terms—is very difficult in relative terms. This signals the need for greater funding of AV ethics that seeks to develop concrete solutions for OEMs and policymakers.

4 Conclusions

This chapter outlined a series of works produced as part of the 2019 Autonomous Vehicles Symposium, and reflections on the private discussion within that breakout session. The session included varied presentations on the ethics of AV decision algorithms, their relation to other technologies, and their broader implications. Participants, including presenters, were drawn from private industry, academia, and government; and from early career and established practitioners. Observed points of deliberation concerned: the need for "before the crash" ethical algorithmic decision-making; concrete and robust ethical theories of AV action; a focus on the larger context for AVs; and better resources and financing for the development of ethics and AVs.

This provides a program for further work for a variety of stakeholders. For researchers and practitioners engaged with AV ethics, it provides two central calls to action to better develop a range of ethical decision-making tools, including those that anticipate ethical dilemmas; and motivate a more serious treatment of ethics beyond canvassing issues. For practitioners and policymakers, it provides a guide to expand the sphere of the conversation around the ethics of AVs. And for funders—private or public—it invites the creation of streams of funding for interdisciplinary research into the ethics of AVs that favors collaboration between empirical and conceptual researchers, and decision-makers.

Acknowledgement. Many thanks to Morgan Avera and Pamela Robinson for their work in organizing the workshop and taking minutes during the proceedings, on which this report is based.

References

Awad, E., Dsouza, S., Kim, R., et al.: The moral machine experiment. Nature **563**, 59–64 (2018). https://doi.org/10.1038/s41586-018-0637-6

Bonnefon, J.F., Shariff, A., Rahwan, I.: The social dilemma of autonomous vehicles. Science **352**, 1573–1576 (2016). https://doi.org/10.1126/science.aaf2654

Goodall, N.: Ethical decision making during automated vehicle crashes. Transp. Res. Rec. J. Transp. Res. Board **2424**, 58–65 (2014a). https://doi.org/10.3141/2424-07

Goodall, N.J.: Can you program ethics into a self-driving car? IEEE Spect. **53**, 28–58 (2016a). https://doi.org/10.1109/mspec.2016.7473149

Goodall, N.J.: Away from trolley problems and toward risk management. Appl. Artif. Intell. **30**, 810–821 (2016b). https://doi.org/10.1080/08839514.2016.1229922

Goodall, N.J.: Machine ethics and automated vehiclesvehicles. In: Meyer, G., Beiker, S. (eds.) Road Vehicle Automation, pp. 93–102. Springer, Dordecht (2014b)

Goodall, N.J., Santoni Di Sio, F., Mecacci, G., et al.: Ethical and Social Implications of Autonomous Vehicles (2018)

Jiang, L., Xie, Y., Chen, D., Li, T., Evans, N.G.: Dampen the stop-and-go traffic with connected and automated vehicles – a deep reinforcement learning approach. arXiv:2005.08245 [Cs, Eess], 17 May 2020

Lin, P.: Why Ethics Matters for Autonomous Cars. In: Maurer, M., Gerdes, J., Lenz, B., Winner, H. (eds.) Autonomes Fahren, pp. 69–85. Springer, Heidelberg (2015)

Author Index

G. Meyer and S. Beiker (Eds.): AVS 2019, LNMOB, pp. 171–172, 2020.
https://doi.org/10.1007/978-3-030-52840-9

The manufacturer's authorised representative in the EU is Springer Nature Customer Service Centre GmbH, Europaplatz 3, 69115 Heidelberg, Germany. If you have any concerns regarding our products, please contact ProductSafety@springernature.com

Printed and bound by CPI Group (UK) Ltd, Croydon, CR0 4YY
05/10/2026
02228156-0005